Tucholsky   Wagner   Zola   Scott   Sydow   Freud   Schlegel
Turgenev   Wallace   Fonatne
Twain   Walther von der Vogelweide   Fouqué   Friedrich II. von Preußen
Weber   Freiligrath   Frey
Fechner   Fichte   Weiße Rose   von Fallersleben   Kant   Ernst   Frommel
Richthofen
Hölderlin
Engels   Fielding   Eichendorff   Tacitus   Dumas
Fehrs   Faber   Flaubert
Eliasberg   Ebner Eschenbach
Feuerbach   Maximilian I. von Habsburg   Fock   Eliot   Zweig
Ewald   Vergil
Goethe   Elisabeth von Österreich   London
Mendelssohn   Balzac   Shakespeare   Dostojewski   Ganghofer
Lichtenberg   Rathenau   Doyle   Gjellerup
Trackl   Stevenson   Hambruch
Mommsen   Tolstoi   Lenz   Hanrieder   Droste-Hülshoff
Thoma   von Arnim   Hägele   Hauff   Humboldt
Dach   Verne
Reuter   Rousseau   Hagen   Hauptmann   Gautier
Karrillon   Garschin   Defoe   Hebbel   Baudelaire
Damaschke   Descartes
Wolfram von Eschenbach   Dickens   Schopenhauer   Hegel   Kussmaul   Herder
Darwin   Melville   Grimm   Jerome   Rilke   George
Bronner   Bebel   Proust
Campe   Horváth   Aristoteles   Voltaire   Federer
Bismarck   Vigny   Barlach   Heine   Herodot
Gengenbach
Storm   Casanova   Tersteegen   Gilm   Grillparzer   Georgy
Chamberlain   Lessing   Langbein   Gryphius
Brentano   Lafontaine
Strachwitz   Claudius   Schiller   Kralik   Iffland   Sokrates
Katharina II. von Rußland   Bellamy   Schilling
Gerstäcker   Raabe   Gibbon   Tschechow
Löns   Hesse   Hoffmann   Gogol   Wilde   Gleim   Vulpius
Luther   Heym   Hofmannsthal   Klee   Hölty   Morgenstern   Goedicke
Roth   Heyse   Klopstock   Puschkin   Homer   Kleist
Luxemburg   La Roche   Horaz   Mörike   Musil
Machiavelli   Kierkegaard   Kraft   Kraus
Navarra   Aurel   Musset   Moltke
Nestroy   Marie de France   Lamprecht   Kind   Kirchhoff   Hugo
Laotse   Ipsen   Liebknecht
Nietzsche   Nansen   Ringelnatz
Marx   Lassalle   Gorki   Klett   Leibniz
von Ossietzky   May   vom Stein   Lawrence   Irving
Petalozzi   Knigge
Platon   Pückler   Michelangelo   Kock   Kafka
Sachs   Poe   Liebermann   Korolenko
de Sade   Praetorius   Mistral   Zetkin

The publishing house tradition has created the series **TREDITION CLASSICS**. It contains classical literature works from over two thousand years. Most of these titles have been out of print and off the bookstore shelves for decades.

The book series is intended to preserve the cultural legacy and to promote the timeless works of classical literature. As a reader of a **TREDITION CLASSICS** book, the reader supports the mission to save many of the amazing works of world literature from oblivion.

The symbol of **TREDITION CLASSICS** is Johannes Gutenberg (1400 – 1468), the inventor of movable type printing.

With the series, tradition intends to make thousands of international literature classics available in printed format again – worldwide.

All books are available at book retailers worldwide in paperback and in hardcover. For more information please visit: www.tredition.com

tradition was established in 2006 by Sandra Latusseck and Soenke Schulz. Based in Hamburg, Germany, tradition offers publishing solutions to authors and publishing houses, combined with worldwide distribution of printed and digital book content. tradition is uniquely positioned to enable authors and publishing houses to create books on their own terms and without conventional manufacturing risks.

For more information please visit: www.tredition.com

Foods That Will Win The War And How To Cook Them (1918)

C. Houston (Charles Houston) Goudiss

Imprint

This book is part of the TREDITION CLASSICS series.

Author: C. Houston (Charles Houston) Goudiss
Cover design: toepferschumann, Berlin (Germany)

Publisher: tredition GmbH, Hamburg (Germany)
ISBN: 978-3-8495-1011-4

www.tredition.com
www.tredition.de

Copyright:
The content of this book is sourced from the public domain.

The intention of the TREDITION CLASSICS series is to make world literature in the public domain available in printed format. Literary enthusiasts and organizations worldwide have scanned and digitally edited the original texts. tredition has subsequently formatted and redesigned the content into a modern reading layout. Therefore, we cannot guarantee the exact reproduction of the original format of a particular historic edition. Please also note that no modifications have been made to the spelling, therefore it may differ from the orthography used today.

SAVE

1. **wheat** — *use more corn*
2. **meat** — *use more fish & beans*
3. **fats** — *use just enough*
4. **sugar** — *use syrups*

and serve the cause of freedom

U.S. FOOD ADMINISTRATION

food

1 - buy it with thought
2 - cook it with care
3 - serve just enough
4 - save what will keep
5 - eat what would spoil
6 - home-grown is best

don't waste it

[pg 4]

FOREWORD

Food will win the war, and the nation whose food resources are best conserved will be the victor. This is the truth that our government is trying to drive home to every man, woman and child in America. We have always been happy in the fact that ours was the richest nation in the world, possessing unlimited supplies of food, fuel, energy and ability; but rich as these resources are they will not meet the present food shortage unless every family and every individual enthusiastically co-operates in the national saving campaign as outlined by the United States Food Administration.

The regulations prescribed for this saving campaign are simple and easy of application. Our government does not ask us to give up three square meals a day—nor even one. All it asks is that we substitute as far as possible corn and other cereals for wheat, reduce a little our meat consumption and save sugar and fats by careful utilization of these products.

There are few housekeepers who are not eager to help in this saving campaign, and there are few indeed who do not feel the need of conserving family resources. But just how is sometimes a difficult task.

This book is planned to solve the housekeeper's problem. It shows how to substitute cereals and other grains for wheat, how to cut down the meat bill by the use of meat extension and meat substitute dishes which supply equivalent nutrition at much less cost; it shows the use of syrup and other products that save sugar, and it explains [pg 5] how to utilize all kinds of fats. It contains 47 recipes for the making of war breads; 64 recipes on low-cost meat dishes and meat substitutes; 54 recipes for sugarless desserts; menus for meatless and wheatless days, methods of purchasing—in all some two hundred ways of meeting present food conditions at minimum cost and without the sacrifice of nutrition.

Not only have its authors planned to help the woman in the home, conserve the family income, but to encourage those saving habits which must be acquired by this nation if we are to secure a permanent peace that will insure the world against another onslaught by the Prussian military powers.

A little bit of saving in food means a tremendous aggregate total, when 100,000,000 people are doing the saving. One wheatless meal a day would not mean hardship; there are always corn and other products to be used. Yet one wheatless meal a day in every family would mean a saving of 90,000,000 bushels of wheat, which totals 5,400,000,000 lbs. Two meatless days a week would mean a saving of 2,200,000 lbs. of meat per annum. One teaspoonful of sugar per person saved each day would insure a supply ample to take care of our soldiers and our Allies. These quantities mean but a small individual sacrifice, but when multiplied by our vast population they will immeasurably aid and encourage the men who are giving their lives to the noble cause of humanity on which our nation has embarked.

The Authors.

[pg 6]

CONTENTS

FOREWORD

SAVE WHEAT: Reasons Why Our Government Asks Us to Save Wheat, with Practical Recipes for the Use of Other Grains

A General rule for proportions in bread-making

Use of Corn

Use of Oats

Use of Rye

Use of Barley

Use of Potatoes

Use of Mixed Grains

Pancakes and Waffles

SAVE MEAT: Reasons Why Our Government Has Asked Us to Save Meat, with Practical Recipes for Meat Conservation

Selection of Meat , , ,

Methods of Cooking ,

Charts ,

Comparative Composition of Meat and Meat Substitutes

Economy of Meat and Meat Substitutes

Meat Economy Dishes

Fish as a Meat Substitute

Fish Recipes

Cheese as a Meat Substitute

Meat Substitute Dishes

SAVE SUGAR: Reasons Why Our Government Asks Us to Save Sugar, with Practical Recipes for Sugarless Desserts, Cakes, Candies and Preserves

Sugarless Desserts

Sugarless Preserves

SAVE FAT: Reasons Why Our Government Asks Us to Save Fat, with Practical Recipes for Fat Conservation

To Render Fats

Various Uses for Leftover Fats

SAVE FOOD: Reasons Why Our Government Asks Us Not to Waste Food, with Practical Recipes for the Use of Leftovers

A Simple Way to Plan a Balanced Ration

Table Showing Number of Calories per Day Required by Various Classes

Sauces Make Leftovers Attractive

Use of Gelatine in Combining Leftovers

Salads Provide an Easy Method of Using Leftovers

Use of Stale Bread, Cake and Leftover Cereals

Soups Utilize Leftovers

All-in-one-dish Meals — Needing only fruit or simple dessert, bread and butter to complete a well-balanced menu

Wheatless Day Menus

Meatless Day Menus

Meat Substitute Dinners

Vegetable Dinners

Save and Serve — Bread; Meat; Sugar; Fat; Milk; Vegetables ,

Blank Pages for Recording Favorite Family Recipes

The Recipes in this book have been examined and approved by the United States Food Administration
Illustrations furnished by courtesy of the United States Food Administration

All the recipes in this book have been prepared and used in The School of Modern Cookery conducted by *The Forecast Magazine* and have been endorsed by the U.S. Food Administration. They have been worked out under the direction of Grace E. Frysinger, graduate in Domestic Science of Drexel Institute, of Philadelphia, and the University of Chicago. Miss Frysinger, who has had nine years' experience as a teacher of Domestic Science, has earnestly used her skill to make these recipes practical for home use, and at the same time accurate and scientific.

The above illustration shows a class at the School of Modern Cookery. These classes are entirely free, the instruction being given in the interest of household economics. The foods cooked during the demonstration are sampled by the students and in this way it is possible to get in close touch with the needs of the homemakers and the tastes of the average family.

[pg 10]

FOODS THAT WILL WIN THE WAR

[pg 11]

SAVE WHEAT

Reasons Why Our Government Asks Us to Save Wheat, with Practical Recipes for the Use of Other Grains

A slice of bread seems an unimportant thing. Yet one good-sized slice of bread weighs an ounce. It contains almost three-fourths of an ounce of flour.

If every one of the country's 20,000,000 homes wastes on the average only one such slice of bread a day, the country is throwing away daily over 14,000,000 ounces of flour—over 875,000 pounds, or enough flour for over a million one-pound loaves a day. For a full year at this rate there would be a waste of over 319,000,000 pounds of flour—1,500,000 barrels—enough flour to make 365,000,000 loaves.

As it takes four and one-half bushels of wheat to make a barrel of ordinary flour, this waste would represent the flour from over 7,000,000 bushels of wheat. Fourteen and nine-tenths bushels of wheat on the average are raised per acre. It would take the product of some 470,000 acres just to provide a single slice of bread to be wasted daily in every home.

[pg 12]

But some one says, "a full slice of bread is not wasted in every home." Very well, make it a daily slice for every four or every ten or every thirty homes—make it a weekly or monthly slice in every home—or make the wasted slice thinner. The waste of flour involved is still appalling. These are figures compiled by government experts, and they should give pause to every housekeeper who permits a slice of bread to be wasted in her home.

Another source of waste of which few of us take account is home-made bread. Sixty per cent. of the bread used in America is made in the home. When one stops to consider how much home-made bread is poorly made, and represents a large waste of flour, yeast and fuel, this housewifely energy is not so commendable. The bread flour used in the home is also in the main wheat flour, and all waste of wheat at the present time increases the shortage of this most necessary food.

Fuel, too, is a serious national problem, and all coal used in either range, gas, or electric oven for the baking of poor bread is an actual national loss. There must be no waste in poor baking or from poor care after the bread is made, or from the waste of a crust or crumb.

Waste in your kitchen means starvation in some other kitchen across the sea. Our Allies are asking for 450,000,000 bushels of wheat, and we are told that even then theirs will be a privation loaf. Crop shortage and unusual demand has left Canada and the United States, which are the largest sources of wheat, with but 300,000,000 bushels available for export. The deficit must be met by reducing consumption on this side the Atlantic. This can be done by eliminating waste and by making use of cereals and flours other than wheat in bread-making.

The wide use of wheat flour for bread-making has been due to custom. In Europe rye and oats form the staple breads of many countries, and in some sections of the [pg 13] South corn-bread is the staff of life. We have only to modify a little our bread-eating habits in order to meet the present need. Other cereals can well be used to eke out the wheat, but they require slightly different handling.

In making yeast breads, the essential ingredient is gluten, which is extended by carbon dioxide gas formed by yeast growth. With the exception of rye, grains other than wheat do not contain sufficient gluten for yeast bread, and it is necessary to use a wheat in varying proportions in order to supply the deficient gluten. Even the baker's rye loaf is usually made of one-half rye and one-half wheat. This is the safest proportion for home use in order to secure a good texture.

When oatmeal is used, it is necessary to scald the oatmeal to prevent a raw taste. Oatmeal also makes a softer dough than wheat, and it is best to make the loaf smaller and bake it longer: about one hour instead of the forty-five minutes which we allow for wheat bread.

The addition of one-third barley flour to wheat flour makes a light colored, good flavored bread. If a larger proportion than this is used, the loaf has a decided barley flavor. If you like this flavor and increase the proportion of barley, be sure to allow the dough a little longer time to rise, as by increasing the barley you weaken the gluten content of your loaf.

Rice and cornmeal can be added to wheat breads in a 10 per cent. proportion. Laboratory tests have shown that any greater proportion than this produces a heavy, small loaf.

Potato flour or mashed potato can be used to extend the wheat, it being possible to work in almost 50 per cent. of potato, but this makes a darker and moister loaf than when wheat alone is used. In order to take care of this [pg 14] moisture, it is best to reserve part of the wheat for the second kneading.

Graham and entire wheat flour also effect a saving of wheat because a larger percentage of the wheat berry is used. Graham flour is the whole kernel of wheat, ground. Entire wheat flour is the flour resulting from the grinding of all but the outer layer of wheat. A larger use of these coarser flours will therefore help materially in eking out our scant wheat supply as the percentage of the wheat berry used for bread flour is but 72 per cent. Breads made from these coarser flours also aid digestion and are a valuable addition to the dietary.

In order to keep down waste by eliminating the poor batch of bread, it is necessary to understand the principles of bread-making. Fermentation is the basic principle of yeast bread, and fermentation is controlled by temperature. The yeast plant grows at a temperature from 70 to 90 degrees (Fahrenheit), and if care is taken to maintain this temperature during the process of fermentation, waste caused by sour dough or over-fermentation will be eliminated. When we control the temperature we can also reduce the time necessary for making a loaf of bread, or several loaves of bread as may be needed, into as short a period as three hours. This is what is known as the quick method. It not only saves time and labor, but, controlling the temperature, insures accurate results. The easiest way to control the temperature is to put the bowl containing the dough into another of slightly larger size containing water at a temperature of 90 degrees. The water of course should never be hot. Hot water kills the yeast plant. Cold water checks its growth. Cover the bowl and set it in the gas oven or fireless cooker or on the shelf of the coal range. As the water in the large bowl cools off, remove a cupful and add a cupful of hot water. At the end of one and one-half hours the [pg 15] dough should have doubled in bulk. Take it out of the pan and knead until the large gas bubbles are broken (about ten minutes). Then place in greased bread pans and allow to rise for another half hour. At the end of this time it will not only fill the pan, but will project out of it. Do not allow the dough to rise too high, for then the bread will have large holes in it. A good proportion as a general rule to follow, is:

3½ cupfuls of flour (this includes added cereals)

1 cupful of water or milk

½ tablespoon shortening

1½ teaspoons salt

1 cake of compressed yeast

> In this recipe sugar has been omitted because of the serious shortage, but after the war a teaspoon of sugar should be added. The shortening, although small in quantity, may also be omitted.

These materials make a loaf of about one pound, which should be baked in forty to fifty minutes at a temperature of 450 degrees (Fahrenheit). Allow a little longer time for bread containing oatmeal or other grains. Such breads require a little longer baking and a little lower temperature than wheat breads. If you do not use a thermometer in testing your oven, place a piece of paper on the center shelf, and if it browns in two minutes your oven is right. If a longer period for raising is allowed than is suggested in the above recipe, the yeast proportion should be decreased. For overnight bread use one-quarter yeast cake per loaf; for six-hour bread, use one-half yeast cake per loaf; for three-hour bread, use one yeast cake per loaf. In baking, the time allowed should depend on the size of the loaf. When baked at a temperature of 450 degrees, large loaves take from forty-five to sixty minutes, small loaves from thirty to forty minutes, rolls from ten to twenty minutes.

It is well to divide the oven time into four parts. During the first quarter, the rising continues; second quarter, browning begins; the third quarter, browning is finished; [pg 16] the fourth quarter, bread shrinks from the side of the pan. These are always safe tests to follow in your baking. When baked, the bread should be turned out of the pans and allow to cool on a wire rack. When cool, put the bread in a stone crock or bread box. To prevent staleness, keep the old bread away from the fresh—scald the bread crock or give your bread box a sun bath at frequent intervals.

Even with all possible care to prevent waste, yeast breads will not conserve our wheat supply so well as quick breads, because all yeast breads need a larger percentage of wheat. The home baker can better serve her country by introducing into her menus numerous quick breads that can be made from cornmeal, rye, corn and rye, hominy, and buckwheat. Griddle cakes and waffles can also be made from lentils, soy beans, potatoes, rice and peas.

Do not expect that the use of other cereals in bread-making will reduce the cost of your bread. That is not the object. Saving of wheat for war needs is the thing we are striving for, and this is as much an act of loyalty as buying Liberty Bonds. It is to meet the crucial world need of bread that we are learning to substitute, and not to spare the national purse.

Besides this saving of wheat, our Government also asks us to omit all fat from our yeast breads in order to conserve the diminishing fat supply. This may seem impossible to the woman who has never made bread without shortening, but recent experiments in bread-making laboratories have proved that bread, without shortening, is just as light and as good in texture as that made with shortening—the only difference being a slight change in flavor. These experiments have also shown that it is possible to supply shortening by the introduction of 3 per cent. to 5 per cent. of canned cocoanut or of peanut butter, [pg 17] and that sugar may also be omitted from bread-making recipes. In fact, the war is bringing about manifold interesting experiments which prove that edible and nutritious bread can be made of many things besides the usual white flour.

The recipes herewith appended, showing the use of combinations of cereals and wheat, have been carefully tested in The Forecast

School of Modern Cookery. Good bread can be made from each recipe, and the new flavors obtained by the use of other grains make a pleasing and wholesome variety.

A family which has eaten oatmeal or entire wheat bread will never again be satisfied with a diet that includes only bread made from bleached flour. Children, especially, will be benefited by the change, as the breads made from coarser flours are not only more nutritious, but are rich in the minerals and vitamine elements that are so essential to the growth of strong teeth, bones and growing tissues.

The homemaker, too, will never regret her larger acquaintance with bread-making materials, as the greater variety of breads that she will find herself able to produce will be a source of pleasure and keen satisfaction.

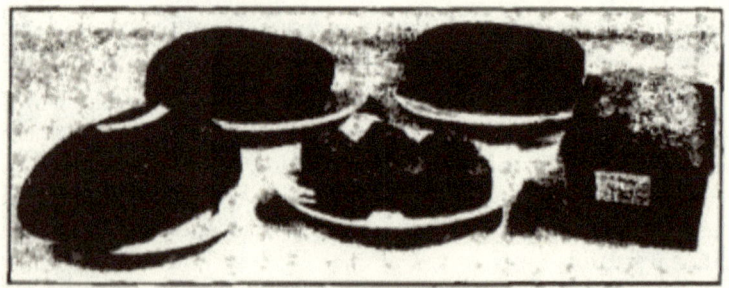

Breads Made From the Coarser Flours, Whole Wheat, Cornmeal, Rye, Conserve Our Wheat Supply
[pg 18]
To Conform to U.S. Food Administration Regulations During the War, Eliminate Fat and Sweetening in Breads — Whenever Fat Is Used, Use Drippings

THE USE OF CORN

CORNMEAL ROLLS

    1 cup bread flour

    1 cup cornmeal

    4 teaspoons baking powder

    2 tablespoons fat

    1 egg

    ⅓ cup milk

    1½ teaspoons salt

    1 tablespoon sugar

Mix and sift dry ingredients and cut in the fat. Beat the egg and add to it the milk. Combine the liquid with the dry ingredients. Shape as Parker House rolls and bake in a hot oven 12 to 15 minutes.

BUTTERMILK OR SOUR MILK CORNMEAL MUFFINS

    2 cups cornmeal

    1 egg

    2 tablespoons sugar

    2 tablespoons fat

    2 cups sour or buttermilk

    1 teaspoon salt

    1 teaspoon soda

Dissolve soda in a little cold water. Mix ingredients adding soda last. Bake in hot oven 20 minutes.

CORNMEAL GRIDDLE CAKES

1⅓ cups cornmeal

1½ cups boiling water

¾ cup milk

2 tablespoons fat

1 tablespoon molasses

⅔ cup flour

1½ teaspoons salt

4 teaspoons baking powder

Scald meal with boiling water. Add milk, fat and molasses. Add sifted dry ingredients. Bake on hot griddle.

SOUTHERN SPOON BREAD

1 cup white cornmeal

2 cups boiling water

¼ cup bacon fat or drippings

3 teaspoons baking powder

1 teaspoon salt

2 eggs

3 slices bread

½ cup cold water

1 cup milk

[pg 19]

Scald cornmeal with boiling water. Soak bread in cold water and milk. Separate yolks and whites of eggs. Beat each until light. Mix ingredients in order given, folding in whites of eggs last. Bake in buttered dish in hot oven 50 minutes.

SPOON BREAD

    2 cups water

    1 cup milk

    1 cup cornmeal

    ⅓ cup sweet pepper

    1 tablespoon fat

    2 eggs

    2 teaspoons salt

Mix water and cornmeal and bring to the boiling point and cook 5 minutes. Beat eggs well and add with other materials to the mush. Beat well and bake in a well-greased pan for 25 minutes in a hot oven. Serve from the same dish with a spoon. Serve with milk or syrup.

CORNMEAL RAGGED ROBINS

    1½ cups cornmeal

    1 cup bread flour

    1½ teaspoons salt

    1⅓ cups milk

    2½ teaspoons cream of tartar

    4 tablespoons fat

    1¼ teaspoons soda

Sift dry ingredients. Cut in the fat. Add liquid and drop by spoonfuls on greased baking sheet. Bake in hot oven 12 to 15 minutes. These may be rolled and cut same as baking powder biscuits.

INDIAN PUDDING

    4 cups milk

⅓ cup cornmeal

⅓ cup molasses

1 teaspoon salt

1 teaspoon ginger

1 teaspoon allspice

Cook milk and meal in a double boiler 20 minutes; add molasses, salt and ginger. Pour into greased pudding dish and bake two hours in a slow oven, or use fireless cooker. Serve with milk. This makes a good and nourishing dessert. Serves six.

TAMALE PIE

2 cups cornmeal

5 cups water (boiling)

2 tablespoons fat

1 teaspoon salt

1 onion

2 cups tomatoes

2 cups cooked or raw meat cut in small pieces

¼ cup green peppers

To the cornmeal and 1 teaspoon salt, add boiling water. Cook one-half hour. Brown onion in fat, add meat. Add salt, ⅛ teaspoon cayenne, the tomatoes and green peppers. Grease baking dish, put in layer of cornmeal mush, add seasoned meat, and cover with mush. Bake one-half hour.

[pg 20]

EGGLESS CORN BREAD

1 cup cornmeal

½ cup bread flour

3 tablespoons molasses

1 cup milk

3 teaspoons baking powder

2 teaspoons salt

2 tablespoons fat

Beat thoroughly. Bake in greased muffin pans 20 minutes.

SWEET MILK CORN BREAD

2 cups cornmeal

2 cups sweet milk (whole or skim)

4 teaspoons baking powder

2 tablespoons corn syrup

2 tablespoons fat

1 teaspoon salt

1 egg

Mix dry ingredients. Add milk, well-beaten egg, and melted fat. Beat well. Bake in shallow pan for about 30 minutes.

SOUR MILK CORN BREAD

2 cups cornmeal

2 cups sour milk

1 teaspoon soda

2 tablespoons fat

2 tablespoons corn syrup or molasses

1 teaspoon salt

1 egg

Mix dry ingredients. Add milk, egg and fat. Beat well. Bake in greased pan 20 minutes.

THE USE OF OATS

COOKED OATMEAL BREAD

>3 cups thick cooked oatmeal
>
>2 tablespoons fat
>
>1½ tablespoons salt
>
>3 tablespoons molasses
>
>1½ cakes yeast
>
>¾ cup lukewarm water
>
>About 5 cups flour

To oatmeal add the sugar, salt and fat. Mix the yeast cake with the lukewarm water, add it to the other materials and stir in the flour until the dough will not stick to the sides of the bowl. Knead until elastic, ten to fifteen minutes, moisten the top of the dough with a little water to prevent a hard crust forming, and set to rise in a warm place. When double its bulk, knead again for a few minutes. Shape into loaves and put into greased pans. Let rise double in bulk and bake in a moderate oven for about 50 minutes.

[pg 21]

OATMEAL BREAD

>2 cups rolled oats
>
>2 cups boiling water
>
>⅓ cup molasses
>
>1 yeast cake
>
>¾ cup lukewarm water
>
>1 tablespoon salt

2 tablespoons fat (melted)

About 6 cups bread flour

Scald the rolled oats with the boiling water and let stand until cool. Dissolve the yeast in the lukewarm water and add to the first mixture when cool. Add the molasses, salt and melted fat. Stir in enough bread flour to knead. Turn on a floured board. Knead lightly. Return to bowl and let rise until double in bulk. Knead and shape in loaves and let rise until double again. Bake in a moderate oven 45 minutes.

OATMEAL NUT BREAD

1 cake compressed yeast

2 cups boiling water

1½ cup lukewarm water

2 cups rolled oats

1 teaspoon salt

¼ cup brown sugar or 2 tablespoons corn syrup

2 tablespoons fat

4 cups flour

½ cup chopped nuts.

Pour two cups of boiling water over oatmeal, cover and let stand until lukewarm. Dissolve yeast and sugar in one-half cup lukewarm water, add shortening and add this to the oatmeal and water. Add one cup of flour, or enough to make an ordinary sponge. Beat well. Cover and set aside in a moderately warm place to rise for one hour.

Add enough flour to make a dough—about three cups, add nuts and the salt. Knead well. Place in greased bowl, cover and let rise in a moderately warm place until double in bulk—about one and one-half hour. Mould into loaves, fill well-greased pans half full, cover and let rise again one hour. Bake forty-five minutes in a moderate oven.

OATMEAL SCONES

>1 cup cold porridge (stiff)
>
>1 cup boiling water
>
>1 tablespoon fat
>
>½ teaspoon baking powder or ¼ teaspoon soda
>
>1 teaspoon corn syrup
>
>½ teaspoon salt

Mix soda, boiling water and fat. Mix all. Turn on board. Mould flat — cut ¼-inch thick and bake on griddle.

OATMEAL MUFFINS

>1⅓ cups flour
>
>2 tablespoons molasses
>
>½ teaspoon salt
>
>2 tablespoons fat
>
>3 teaspoons baking powder
>
>1 egg beaten
>
>½ cup milk
>
>1 cup cooked oatmeal

[pg 22]

Sift dry ingredients. Add egg and milk. Add fat and cereal. Beat well. Bake in greased tins 20 minutes.

ROLLED OATS RAGGED ROBINS

>1½ cups rolled oats

    1 cup bread flour

    1⅓ teaspoons salt

    1⅓ cups milk

    2½ teaspoons cream of tartar

    4 tablespoons fat

    1¼ teaspoons soda

Sift dry ingredients. Cut in the fat. Add liquid and drop by spoonfuls on greased baking sheet. Bake in hot oven 12 to 15 minutes. These may be rolled and cut same as baking powder biscuits. (If uncooked rolled oats are used, allow to stand in the milk for 30 minutes before making recipe.)

THE USE OF RYE

RYE YEAST BREAD

    1 cup milk and water, or water

    1 tablespoon fat

    2 tablespoons corn syrup

    1 teaspoon salt

    2½ cups rye flour

    2½ cups wheat flour

    ½ cake compressed yeast

    2 tablespoons water

Combine ingredients. Mix into dough and knead. Let rise until double original bulk. Knead again. When double bulk, bake about

RYE ROLLS

    4 cups rye flour

1½ teaspoons salt

6 teaspoons baking powder

1½ cups milk

2 tablespoons fat

1 cup chopped nuts

Mix dry ingredients thoroughly. Add milk, nuts and melted shortening. Knead. Shape into rolls. Put into greased pans. Let stand one-half hour. Bake in moderate oven 30 minutes.

WAR BREAD

2 cups boiling water

2 tablespoons sugar

1½ teaspoons salt

¼ cup lukewarm water

2 tablespoons fat

6 cups rye flour

1½ cups whole wheat flour

1 cake yeast

To the boiling water, add the sugar, fat and salt. When lukewarm, add the yeast which has been dissolved in the lukewarm water. Add the rye and whole wheat flour. Cover and let rise until twice its bulk, shape into loaves; let rise until double and bake about 40 minutes, in a moderately hot oven.

[pg 23]

RYE RAGGED ROBINS

1½ cups rye flour

1 cup bread flour

    1½ teaspoons salt

    1⅓ cups milk

    2½ teaspoons cream of tartar

    4 tablespoons fat

    1¼ teaspoons soda

Sift dry ingredients. Cut in the fat. Add liquid and drop by spoonfuls on greased baking sheet. Bake in hot oven 12 to 15 minutes. These may be rolled and cut same as baking powder biscuits.

THE USE OF BARLEY

BARLEY YEAST BREAD

    1 cup milk and water, or water

    2 tablespoons corn syrup

    1 tablespoon fat

    1½ teaspoons salt

    1⅙ cups barley flour

    2⅓ cups wheat flour

    ½ cake compressed yeast

Soften the yeast in ¼ cup lukewarm liquid. Combine ingredients. Mix into a dough. Knead and let rise to double original bulk. Knead again. Put in pan; when again double in bulk bake 45 minutes.

BARLEY MUFFINS

    1¼ cups whole wheat flour

    1 cup barley meal

    ½ teaspoon salt

3 teaspoons baking powder

1 egg

1¼ cups sour milk

½ teaspoon soda

2 tablespoons drippings

Sift flour, barley meal, salt and baking powder. Dissolve soda in a little cold water and add to sour milk. Combine flour mixture and sour milk, add beaten egg and melted fat. Bake in muffin pans in a moderate oven 25 minutes.

BARLEY SPOON BREAD

2 tablespoons pork drippings

3 cups boiling water

1 cup barley meal

2 eggs

Heat drippings in saucepan until slightly brown, add water and when boiling, add barley meal, stirring constantly. Cook in a double boiler one-half hour, cool, and add well-beaten yolks. Fold in whites, beaten. Bake in greased dish in moderate oven one-half hour.

[pg 24]

BARLEY PUDDING

5 cups milk

½ cup barley meal

½ teaspoon salt

½ teaspoon ginger

¾ cup molasses

Scald the milk, pour this on the meal and cook in double boiler one-half hour; add molasses, salt and ginger. Pour into greased pudding dish and bake two hours in a slow oven. Serve either hot or cold with syrup.

BARLEY SCONES

    1 cup whole wheat flour

    1 cup barley meal

    ½ teaspoon salt

    2 teaspoons baking powder

    3 tablespoons fat

    ¾ cup sour milk

    ⅓ teaspoon soda

Sift flour, barley meal, salt and baking powder together. Add fat. Dissolve soda in one tablespoon cold water and add to sour milk. Combine flour mixture and sour milk to form a soft dough. Turn out on a well-floured board, knead slightly, roll to one-half inch thickness; cut in small pieces and bake in a hot oven 15 minutes.

THE USE OF POTATO

POTATO BISCUIT

    1 cup mashed lightly packed potato

    2 tablespoons fat

    1 cup whole wheat flour

    1 teaspoons baking powder

    1 teaspoon salt

    About ½ cup milk or water in which potatoes were cooked

Add melted fat to mashed potato. Mix and sift flour, baking powder and salt and add to potato mixture, add enough of the milk to make a soft dough. Roll out ½ inch thick, cut with a biscuit cutter and bake in a quick oven for 15 minutes. (If bread flour is used in place of whole wheat, the biscuits are slightly lighter and flakier in texture.)

POTATO BREAD

    1½ cups tightly packed mashed potato

    2½ cups wheat flour

    1 tablespoon warm water

    ½ yeast cake

    ½ teaspoon salt

Make dough as usual. Let rise in warm place for 15 minutes. Mould into loaf, put in pan, let rise until double in bulk in warm place. Bake for 45 minutes in hot oven.

[pg 25]

POTATO YEAST BREAD

    ½ cup milk and water or water

    2 tablespoons corn syrup

    4 tablespoons fat

    1½ teaspoons salt

    4 cups boiled potatoes

    8 cups flour

    ½ cake compressed yeast

    ¼ cup warm water

Dissolve yeast in the warm water. Add other ingredients and make same as any bread.

POTATO PARKER HOUSE ROLLS

    ½ cake yeast

    1 cup milk (scalded)

    1 teaspoon fat

    3 tablespoons corn syrup (or 1 tablespoon sugar)

    3½ cups flour

    2 cups potato (mashed and hot)

    1 teaspoon salt

    1 egg

Dissolve yeast in milk (luke warm). Stir in dry ingredients. Add potato and knead until smooth. Let rise until light. Roll thin, fold over, bake until brown.

THE USE OF MIXED GRAINS

WAR BREAD OR THIRDS BREAD

    1 pint milk, or milk and water

    2 teaspoons salt

    2 tablespoons molasses

    1 yeast cake

    2 tablespoons fat

Mix as ordinary bread dough. Add 2 cups cornmeal and 2 cups rye meal and enough whole wheat flour to knead. Let rise, knead, shape, let rise again in the pan and bake 45 minutes.

CORN MEAL AND RYE BREAD

    2 cups lukewarm water

    1 cake yeast

2 teaspoons salt

⅓ cup molasses

1¼ cup rye flour

1 cup corn meal

3 cups bread flour

Dissolve yeast cake in water, add remaining ingredients, and mix thoroughly. Let rise, shape, let rise again and bake.

[pg 26]

BOSTON BROWN BREAD

1 cup rye meal

1 cup cornmeal

1 cup graham flour

2 cups sour milk

1¾ teaspoons soda

1½ teaspoons salt

¾ cup molasses

Beat well. Put in greased covered molds, steam 2 to 3 hours.

BREAD MUFFINS

2 cups bread crumbs

⅓ cup flour

1 tablespoon fat, melted

1½ cups milk

1 egg

2 teaspoons baking powder

½ teaspoon salt

Cover crumbs with milk and soak 10 minutes. Beat smooth, add egg yolks, dry ingredients sifted together and fat. Fold in beaten whites of eggs. Bake in muffin tins in moderate oven for 15 minutes.

CORN, RYE AND WHOLE WHEAT FRUIT MUFFINS

    1/3 cup boiling water

    1 cup cornmeal

    1/4 teaspoon soda

    1/4 cup molasses

    1 cup whole wheat flour

    1 cup rye flour

    3 teaspoons baking powder

    1 teaspoon salt

    1 cup milk

    1/3 cup raisins cut in halves

    1/4 cup chopped nuts

    2 tablespoons fat

Scald meal with boiling water, mix soda and molasses. Mix dry ingredients, mix all thoroughly. Bake in muffin pans one-half hour.

SOY BEAN MEAL BISCUIT

    1 cup soy bean meal or flour

    1 cup whole wheat

    1 1/2 teaspoons salt

    4 teaspoons baking powder

    1 tablespoon corn syrup

    2 tablespoons fat

    1 cup milk

Sift dry ingredients. Cut in fat. Add liquid to make soft dough. Roll one-half inch thick. Cut and bake 12 to 15 minutes in hot oven.

EMERGENCY BISCUIT

    1 cup whole wheat flour

    1 cup cornmeal

    1 tablespoon fat

    ½ teaspoon soda

    1 cup sour milk

    1 teaspoon salt

Mix as baking powder biscuit. Drop by spoonfuls on greased baking sheet. Bake 15 minutes in hot oven.

[pg 27]

PANCAKES AND WAFFLES

SOUR MILK PANCAKES

    1 cup sour milk

    ½ cup cooked cereal or

    1 cup bread crumbs

    1 tablespoon melted fat

    1 egg

    ¾ cup whole wheat flour

    1 teaspoon soda

    ⅛ teaspoon salt

Mix bread crumbs, flour, salt; add beaten egg, fat and cereal; mix soda with sour milk and add to other ingredients.

SPLIT PEA PANCAKES

    2 cups split peas

    2 egg whites

    ⅓ cup flour

    1 cup milk

    2 egg yolks

    2 tablespoons pork drippings

    ⅛ teaspoon cayenne

    1 teaspoon salt

    1 teaspoonful baking powder

Soak peas over night, cook, and when tender, put through a food chopper and mix the ingredients. Bake on hot greased griddle.

BREAD GRIDDLE CAKES

    2 cups sour milk

    2 cups bread

Let stand until soft

Put through colander. For each one pint use:

    1 egg

    1 teaspoon soda

    2 teaspoons sugar

    ¼ teaspoon salt

    ¾ cup flour

    1 egg beaten

Mix well; bake at once on hot greased griddle.

OATMEAL PANCAKES

    2 cups oatmeal

    1 tablespoon melted fat

    1/8 teaspoon salt

Add:

    1 egg beaten into a cupful of milk

    1 cupful flour into which has been sifted 1 teaspoonful baking powder.

Beat well. Cook on a griddle. This is an excellent way to use leftover oatmeal.

[pg 28]

POTATO PANCAKES

    2 cups of chopped potato

    1/2 cup milk

    1 egg

    1 teaspoon salt

    2 cups flour

    5 teaspoons of baking powder

    2 cups of hot water

Parboil potatoes in the skins for fifteen minutes. Pare and chop fine or put through food chopper. Mix potatoes, milk, eggs and salt. Sift the flour and baking powder and stir into a smooth batter. Thin with hot water as necessary. Bake on a greased griddle.

RICE WAFFLES

    1 cup cold boiled rice

1½ cups milk

2 eggs

2 cups flour

⅓ teaspoon salt

1 tablespoon melted fat

4 teaspoons baking powder

Add milk to rice and stir until smooth. Add salt, egg yolks beaten; add flour sifted with baking powder and salt; add fat; add stiffly beaten whites.

RICE GRIDDLE CAKES

½ cup boiled rice

½ cup flour

3 tablespoons fat

1 pint milk

⅔ teaspoon salt

½ teaspoon soda

Stir rice in milk. Let stand one-half hour. Add other ingredients, having dissolved soda in one tablespoon cold water.

CORNMEAL WAFFLES

1 cup cornmeal

½ cup flour

½ teaspoon salt

2 teaspoons baking powder

¼ cup corn syrup

1 egg

1 pint milk

1 tablespoon fat

Cook cornmeal and milk in double boiler 10 minutes. Sift dry ingredients. Add milk, cornmeal; beaten yolks; fat, beaten whites.

CORNMEAL AND RYE WAFFLES

1 cup rye flour

¾ cup cornmeal

1 teaspoon salt

4 teaspoons baking powder

1 tablespoon melted fat

2 eggs

1¼ cups milk

Sift dry ingredients. Add beaten yolks added to milk. Add fat and stiffly beaten whites. If waffles are not crisp add more liquid.

[pg 29]

Each Food Shown is Equivalent in Protein to the Platter of Meat in the Center of the Picture.

SAVE MEAT

Reasons Why Our Government Has Asked Us to Save Meat with Practical Recipes for Meat Conservation

As a nation we eat and waste 80 per cent. more meat than we require to maintain health. This statement, recently issued by the United States Food Administration, is appalling when we consider that there is a greater demand for meat in the world to-day than ever before, coupled with a greatly decreased production. The increase in the demand for meat and animal products is due to the stress of the war. Millions of men are on the fighting line doing hard physical labor, and require a larger food allowance than when they were civilians. To meet the demand for meat and to save their grains, our Allies have been compelled to kill upward of thirty-three million head of their stock animals, and they have thus stifled their animal production. This was burning the [pg 30] candle at both ends, and they now face increased demand handicapped by decreased production.

America must fill the breach. Not only must we meet the present increased demand, but we must be prepared as the war advances to meet an even greater demand for this most necessary food. The way out of this serious situation is first to reduce meat consumption to the amount really needed and then to learn to use other foods that will supply the food element which is found in meat. This element is called protein, and we depend upon it to build and repair body tissues.

Although most persons believe that protein can only be obtained from meat, it is found in many other foods, such as milk, skim milk, cheese, cottage cheese, poultry, eggs, fish, dried peas, beans, cow peas, lentils and nuts. For instance, pound for pound, salmon, either fresh or canned, equals round steak in protein content; cream cheese contains one-quarter more protein and three times as much fat; peanuts (hulled) one-quarter more protein and three and a half times as much fat; beans (dried) a little more protein and one-fifth as much fat; eggs (one dozen) about the same in protein and one-half more fat. It is our manifest duty to learn how to make the best use of these foods in order to save beef, pork and mutton, to be shipped across the sea. This means that the housekeeper has before her the task of training the family palate to accept new food preparations. Training the family palate is not easy, because bodies that have grown accustomed to certain food combinations find it difficult to get along without them, and rebel at a change. If these habits of diet are suddenly disturbed we may upset digestion, as well as create a feeling of dissatisfaction which is equally harmful to physical well-being. The wise housekeeper will therefore make her changes gradually.

In reducing meat in the diet of a family that has been [pg 31] used to having meat twice a day, it will be well to start out with meat once a day and keep up this régime for a couple of weeks. Then drop meat for a whole day, supplying in its stead a meat substitute dish that will furnish the same nutriment. After a while you can use meat substitutes at least twice a week without disturbing the family's mental or physical equilibrium. It would be well also to introduce dishes that extend the meat flavor, such as stews combined with dumplings, hominy, or rice; pot pies or short cakes with a dressing of meat and vegetables; meat loaf, souffle or croquettes in which meat is combined with bread crumbs, potato or rice.

Meat eating is largely a matter of flavor. If flavor is supplied, the reduction of meat in the diet can be made with little annoyance. Nutrition can always be supplied in the other dishes that accompany the meal, as a certain proportion of protein is found in almost every food product. The meat that we use to obtain flavor in sauces and gravies need not be large in quantity, nor expensive in cut. The poor or cheap cuts have generally more flavor than the expensive ones, the difference being entirely in texture and tenderness, freedom from gristle and inedible tissue. There are many cereals, such as rice, hominy, cornmeal, samp and many vegetable dishes, especially dried beans of all kinds, that are greatly improved by the addition of meat sauce and when prepared in this way may be served as the main dish of a meal.

Dr. Harvey W. Wiley has stated that the meat eating of the future will not be regarded as a necessity so much as it has been in the past, and that meat will be used more as a condimental substance. Europe has for years used meat for flavor rather than for nutriment.

It would seem that the time has come for Americans to learn the use of meat for flavor and to utilize more skillfully the protein of other foods.

[pg 32]

It may be difficult to convince the meat lover that he can radically reduce the proportion of meat in his diet without detriment to health. Many persons adhere to the notion that you are not nourished unless you eat meat; that meat foods are absolutely necessary to maintain the body strength. This idea is entirely without foundation, for the foods mentioned as meat substitutes earlier in this chapter can be made to feed the world, and feed it well—in fact, no nation uses so large a proportion of meat as America.

The first step, therefore, in preparing ourselves to reduce meat consumption is to recognize that only a small quantity of meat is necessary to supply sufficient protein for adult life. The growing child or the youth springing into manhood needs a larger percentage of meat than the adult, and in apportioning the family's meat ration this fact should not be overlooked.

The second step is to reduce the amount purchased, choosing cuts that contain the least waste, and by utilizing with care that which we do purchase. Fat, trimmings, and bones all have their uses and should be saved from the garbage pail.

Careful buying, of course, depends on a knowledge of cuts, a study of the percentage of waste in each cut, and the food value of the different kinds of meat. Make a study of the different cuts, as shown in the charts on pages 36, 37, and armed with this knowledge go forth to the butcher for practical buying.

Then comes the cooking, which can only be properly done when the fundamental principles of the cooking processes, such as boiling, braising, broiling, stewing, roasting and frying are understood. Each cut requires different handling to secure the maximum amount of nutriment and flavor. The waste occasioned by improper cooking is a large factor in both household and national economy.

[pg 33]

It has been estimated that a waste of an ounce each day of edible meat or fat in the twenty million American homes amounts to 456,000,000 pounds of valuable animal food a year. At average dressed weights, this amounts to 875,000 steers, or over 3,000,000 hogs. Each housekeeper, therefore, who saves her ounce a day aids in this enormous saving, which will mean so much in the feeding of our men on the fighting line.

So the housekeeper who goes to her task of training the family palate to accept meat substitutes and meat economy dishes, who revolutionizes her methods of cooking so as to utilize even "the pig's squeak," will be doing her bit toward making the world safe for democracy.

The following charts, tables of nutritive values and suggested menus have been arranged to help her do this work. The American woman has her share in this great world struggle, and that is the intelligent conservation of food.

SELECTION OF MEAT

BEEF — Dull red as cut, brighter after exposure to air; lean, well mottled with fat; flesh, firm; fat, yellowish in color. Best beef from animal 3 to 5 years old, weighing 900 to 1,200 pounds. Do not buy wet, soft, or pink beef.

VEAL — Flesh pink. (If white, calf was bled before killed or animal too young.) The fat should be white.

MUTTON — Best from animal 3 years old. Flesh dull red, fat firm and white.

LAMB — (Spring Lamb 3 months to 6 months old; season, February to March.) Bones of lamb should be small; end of bone in leg of lamb should be serrated; flesh pink, and fat white.

PORK — The lean should be fine grained and pale pink. The skin should be smooth and clear. If flesh is soft, or fat yellowish, pork is not good.

[pg 34]

SELECTION OF TOUGHER CUTS AND THEIR USES

Less expensive cuts of meat have more nourishment than the more expensive, and if properly cooked and seasoned, have as much tenderness. Tough cuts, as chuck or top sirloin, may be boned and rolled and then roasted by the same method as tender cuts, the only difference will be that the tougher cuts require longer cooking. Have the bones from rolled meats sent home to use for soups. Corned beef may be selected from flank, naval, plate or brisket. These cuts are more juicy than rump or round cuts.

1. *For pot roast* use chuck, crossrib, round, shoulder, rump or top sirloin.

2. *For stew* use shin, shoulder, top sirloin or neck.

3. *For steaks* use flank, round or chuck. If these cuts are pounded, or both pounded and rubbed with a mixture of 1 part vinegar and 2 parts oil before cooking, they will be very tender.

4. *Soups*—Buy shin or neck. The meat from these may be utilized by serving with horseradish or mustard sauce, or combined with equal amount of fresh meat for meat loaf, scalloped dish, etc.

DRY METHODS

1. *Roasting or Baking*—Oven roasting or baking is applied to roasts.

Place the roast in a hot oven, or if gas is used, put in the broiling oven to sear the outside quickly, and thus keep in the juices. Salt, pepper and flour. If an open roasting pan is used place a few tablespoonfuls of fat and 1 cup of water in the pan, which should be used to baste the roast frequently. If a covered pan is used basting is unnecessary.

Beef or mutton	(5 to 8 lbs.)	10 min. to the lb.	10 min. extra
Lamb	(5 to 8 lbs.)	12 min. to the lb.	12 min. extra
Veal	(5 to 8 lbs.)	15 min. to the lb.	15 min. extra
Pork	(5 to 8 lbs.)	25 min. to the lb.	25 min. extra
Turkey		20 min. to the lb.	
Chicken		30 min. to the lb.	

Duck	30 min. to the lb.
Goose	30 min. to the lb.
Game	30 min. to the lb.

2. *Broiling* — Cooking over or under clear fire. This method is used for chops or steaks.

Sear the meat on both sides. Then reduce the heat and turn the meat frequently. Use no fat.

Time Table — (Count time after meat is seared).

½ inch chops or steaks, 5 minutes

1 inch chops or steaks, 10 minutes

2 inch chops or steaks, 15 to 18 minutes

3. *Pan Broiling* — Cooking in pan with no fat. *Time table same as for broiling* chops, steaks, etc.

4. *Sautéing* — Cooking in pan in small amount of fat. Commonly termed "frying." Used for steaks, chops, etc. *Time table same as for broiling.*

[pg 35]

MOIST METHODS

1. Boiling — Cooking in boiling water — especially poultry, salt meats, etc.

2. Steaming — A method of cooking by utilizing steam from boiling water, which retains more food value than any other. Too seldom applied to meats.

3. Frying — Cooking by immersion in hot fat at temperature 400 to 450 degrees Fahrenheit. Used for croquettes, etc.

If a fat thermometer is not available, test by using small pieces of bread. Put into heated fat:

A — For croquettes made from food requiring little cooking, such as oysters, or from previously cooked mixtures, as rice, fish or meat croquettes, bread should brown in one-half minute.

B—For mixtures requiring cooking, as doughnuts, fritters, etc., bread should brown in one minute.

COMBINATION METHODS

1. Pot Roasting—Cooking (by use of steam from small amount of water) tough cuts of meat which have been browned but not cooked thoroughly.

Season meat. Dredge with flour. Sear in hot pan until well browned. Place oil rack in pot containing water to height of one inch, but do not let water reach the meat. Keep water slowly boiling. Replenish as needed with boiling water. This method renders tough cuts tender, but requires several hours cooking.

2. Stewing—A combination of methods which draws part of flavor into gravy and retains part in pieces which are to be used as meat.

Cut meat into pieces suitable for serving. Cover one-half of meat with cold water. Let stand one hour. Bring slowly to boiling point. Dredge other half of meat with flour and brown in small amount of fat. Add to the other mixture and cook slowly 1½ to 2 hours, or until tender, adding diced vegetables, thickening and seasoning as desired one-half hour before cooking is finished.

3. Fricasseeing—Cooking in a sauce until tender, meat which has been previously browned but not cooked throughout.

Brown meat in small amount of fat. Place in boiling water to cover. Cook slowly until tender. To 1 pint of water in which meat is cooked, add ¼ cup flour, 1 teaspoon salt, ¼ teaspoon cayenne, and ¼ cup milk, thoroughly blended. When at boiling point, add one beaten egg, 1 tablespoon chopped parsley and 1 tablespoon cold water well mixed, Add cooked meat and serve.

[pg 36]

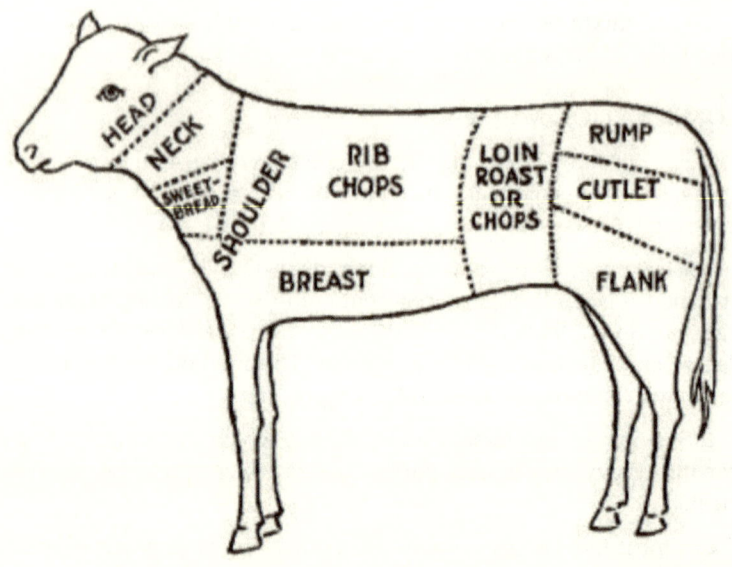

VEAL

Neck for stews.

Shoulder for inexpensive chops.

Sweetbread — broiled or creamed.

Breast for roast or pot roast.

Loin for roast.

Rump for stews.

Cutlet for broiling.

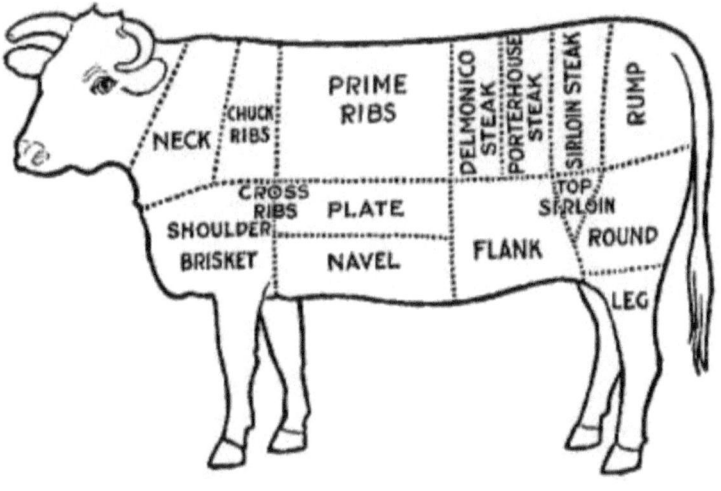

BEEF

[pg 37]

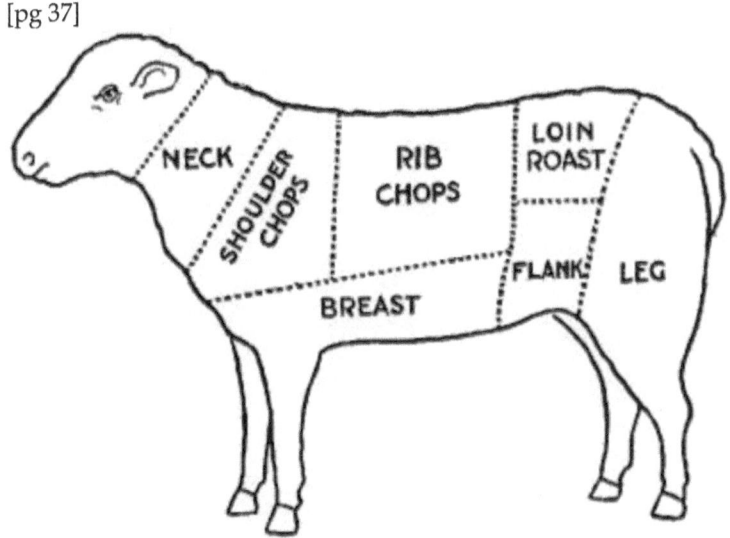

LAMB AND MUTTON

Neck — use for stews.

Shoulder for cheaper chops.

Breast for roast

Ribs for chops or crown roast.

Loin for roast.

Flank for stews.

Leg for cutlet and roast.

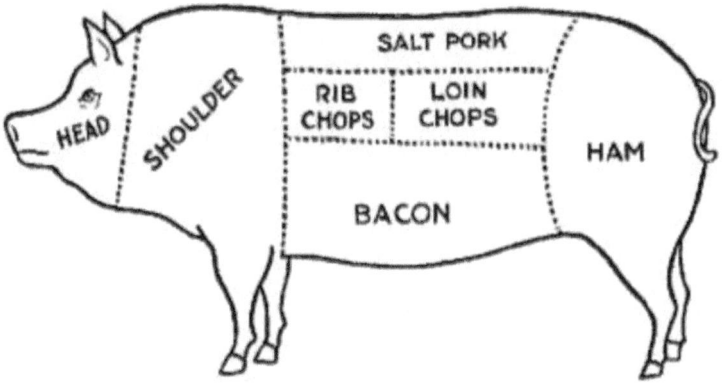

PORK

Head for cheese.

Shoulder same as ham but have it boned. Has same flavor and is much cheaper.

Loin used for chops or roast.

Ham for boiling, roasting or pan broiling.

[pg 38]

LESS-USED EDIBLE PARTS OF ANIMAL, AND METHODS OF COOKING BEST ADAPTED TO THEIR USE

ORGAN	ANIMAL SOURCE	METHODS OF COOKING
Brains	Sheep Pork	Broiled or scrambled with egg
Heart	Veal Pork Beef	Stuffed, baked or broiled
Kidney	Beef Lamb Veal	Stewed or sauted
Liver	Beef Veal Lamb	Fried, boiled, sauted or broiled
Sweetbreads	Young Veal Young Beef	Creamed, broiled
Tail	Beef Pork	Soup or boiled
Tongue	Beef Pork	Boiled, pickled, corned
Tripe	Veal	Broiled or boiled
Fat	All Animals	Fried out for cooking or soap making

| Pigs Feet | Pork | Pickled or boiled or used with meat from head for head cheese |

COMPARATIVE COMPOSITION OF MEAT AND MEAT SUBSTITUTES

Name	Water %	Protein %	Fat %	Carbohydrate %	Mineral Matter %	Calories per lb.
Cheese	34.2	25.2	31.7	2.4	3.8	1,950
Eggs	73.7	13.4	10.5	...	1.0	720
Milk	87.0	3.3	4.0	5.0	0.7	310
Beef	54.8	23.5	20.4	...	1.2	1,300
Cod	58.5	11.1	0.2	...	0.8	209
Salmon	64.0	22.0	12.8	...	1.4	923
Peas	85.3	3.6	0.2	9.8	1.1	252
Baked Beans	68.9	6.9	2.5	19.6	2.1	583
Lentils	15.9	25.1	1.0	56.1	1.1	1,620
Peanuts	9.2	25.8	38.6	24.4	0.2	2,490
String Beans	93.7	1.1	0.1	3.8	1.3	92
Walnuts	2.5	18.4	64.4	13.0	1.7	3,182
Almonds	4.8	21.0	54.9	17.3	2.0	2,940

[pg 39]

THE ECONOMY OF MEAT AND MEAT SUBSTITUTES

Don't buy more than your family actually needs. Study and know what the actual needs are, and you will not make unnecessary expenditures.

Learn what the various cuts of meat are, what they can be used for, and which are best suited to the particular needs of your household.

Study the timeliness of buying certain cuts of meats. There are days when prices are lower than normal.

Always check the butcher's weights by watching him closely or by weighing the goods on scales of your own.

Always buy a definite quantity. Ask what the pound rate is, and note any fractional part of the weight. Don't ask for "ten or twenty cents' worth."

Select your meat or fish personally. There is no doubt that high retail prices are due to the tendency of many housewives to do their buying by telephone or through their servants.

Test the freshness of meat and fish. Staleness of meat and fish is shown by loose and flabby flesh. The gills of fresh fish are red and the fins stiff.

Make all the purchases possible at a public market, if you can walk to it, or if carfare will not make too large an increase in the amount you have set aside for the day's buying.

A food chopper can be made to pay for itself in a short time by the great variety of ways it furnishes of utilizing left-overs.

If possible, buy meat trimmings. They cost 20 cents a pound and can be used in many ways.

Buy the ends of bacon strips. They are just as nutritious as sliced bacon and cost 50 per cent. less.

[pg 40]

Learn to use drippings in place of butter for cooking purposes.

Buy cracked eggs. They cost much less than whole ones and are usually just as good.

Keep a stock pot. Drop into it all left-overs. These make an excellent basis for soup stock.

Don't throw away the heads and bones of fish. Clean them and use them with vegetables for fish chowder or cream of fish soup.

Study attractive ways of serving food. Plain, cheap, dishes can be made appetizing if they look attractive on the table.

Experiment with meat substitutes. Cheese, dried vegetables and the cheaper varieties of fish can supply all the nutriment of meat at a much lower cost.

Don't do your cooking "by guess." If the various ingredients are measured accurately, the dish will taste better and cost less.

Don't buy delicatessen food if you can possibly avoid it. Delicatessen meals cost 15 per cent. more than the same meals cooked at home, and the food is not as nourishing. You pay for the cooking and the rent of the delicatessen store, as well as the proprietor's profit.

Don't pay five or ten cents more a dozen for white eggs in the belief that they are superior to brown eggs. The food value of each is the same. The difference in shell color is due to the breed of hen.

Tell the butcher to give you the trimmings of chicken, i.e., the head, feet, fat and giblets. They make delicious chicken soup. The feet contain gelatine, which gives soup consistency.

Buy a tough, and consequently less expensive, chicken and make it tender by steaming it for three hours before roasting.

[pg 41]

Don't put meat wrapped in paper into the ice-box, as the paper tends to absorb the juices.

Try to find a way to buy at least a part of your meats and eggs direct from the farm. You will get fresher, better food, and if it is sent by parcels post it can usually be delivered to your table for much less than city prices.

MEAT ECONOMY DISHES

MOCK DUCK

> 1 flank steak
>
> 1 teaspoon salt
>
> 1 teaspoon pepper
>
> 1 teaspoon Worcestershire sauce
>
> 1 cup breadcrumbs
>
> 1 tablespoon onion juice
>
> 1 tablespoon chopped parsley

½ teaspoon poultry seasoning

1 pint boiling water

⅓ cup of whole wheat flour

Reserve the water and the flour. Mix other ingredients. Spread on steak. Roll the steak and tie. Roll in the flour. Brown in two tablespoons of fat. Add the water—cover and cook until tender.

BEEF STEW

1 lb. of meat from the neck, cross ribs, shin or knuckles

1 sliced onion

¾ cup carrots

½ cup turnips

1 cup potatoes

1 teaspoon salt

¼ teaspoon pepper

½ cup flour

1 quart water

Soak one-half of the meat, cut in small pieces, in the quart of water for one hour. Heat slowly to boiling point. Season the other half of the meat with salt and pepper. Roll in flour. Brown in three tablespoons of fat with the onion. Add to the soaked meat, which has been brought to the boiling point. Cook one hour or until tender. Add the vegetables, and flour mixed with half cup of cold water. Cook until vegetables are tender.

HAM SOUFFLE

1½ cups breadcrumbs

2 cups scalded milk

1½ cups chopped cooked ham

2 egg yolks

1 tablespoon chopped parsley

1 teaspoon minced onion

½ teaspoon paprika

2 egg whites

[pg 42]

PARSLEY SAUCE

2 tablespoons butter

3 tablespoons flour

1 cup milk

½ teaspoon salt

2 tablespoons chopped parsley

For the soufflé, cook together breadcrumbs and milk for two minutes. Remove from fire, add ham and mix well. Add egg yolks, first beating these well; also the parsley (one tablespoon), onion and paprika. Fold in, last of all, the egg whites whipped to a stiff, dry froth. Turn quickly into a well-greased baking dish and bake in moderate oven for thirty-five minutes, or until firm to the touch; meantime, make the parsley sauce, so that both can be served instantly when the soufflé is done; then it will not fall and grow tough.

For the parsley sauce, melt the butter in saucepan and stir in the flour, stirring until perfectly smooth, then add the milk slowly, stirring constantly; cook until thick, stir in the parsley and salt, and serve at once in a gravy boat.

BATTLE PUDDING

BATTER

1 cup flour

½ cup milk

2 teaspoons baking powder

1 egg

4 tablespoons water

½ teaspoon salt

FILLING

2 cups coarsely chopped cold cooked meat

1 tablespoon drippings

1 medium-sized potato

1 cup stock or hot water

salt and pepper

1 small onion

Any cold meat may be used for this. Cut it into inch pieces. Slice the onion and potato and fry in drippings until onion is slightly browned. Add the meat and stock, or hot water, or dissolve in hot water any left-over meat gravy. Cook all together until potato is soft, but not crumbled; season with the pepper and salt. Thicken with a tablespoon of flour and turn into a pudding dish.

Make a batter by sifting together flour, baking-powder and salt; stir in the egg and milk, mixed with the water. Beat hard until free from lumps, then pour over meat and vegetables in the pudding and bake until brown.

CHINESE MUTTON

1 pint chopped cooked mutton

1 head shredded lettuce

1 can cooked peas

⅛ teaspoon pepper

1 tablespoon fat

1½ cups broth

1 teaspoon of salt

Cook 15 minutes. Serve as a border around rice.

[pg 43]

SHEPHERD'S PIE

2 cups chopped cooked mutton

1 teaspoon salt

⅛ teaspoon pepper

¼ teaspoon curry powder

2 cups hominy

1 cup peas or carrots

½ pint of brown sauce or water

Put meat and vegetables in baking dish. Cover with rice, hominy, or samp, which has been cooked. Bake until brown.

SCALLOPED HAM AND HOMINY

2 cups hominy (cooked)

1 cup chopped cooked ham

⅓ cup fat

⅓ cup flour

1 teaspoon of salt

⅛ teaspoon mustard

⅛ teaspoon cayenne

1 egg

1 cup milk

½ cup water

Melt the fat. Add the dry ingredients and the liquid slowly. When at boiling point, add hominy and ham. Stir in the egg. Place in a baking-dish. Cover with buttered crumbs. Bake until brown.

BEEF LOAF

1 tablespoon lemon juice

1 tablespoon sour pickle

2 teaspoons salt

⅛ teaspoon cayenne

1 teaspoon celery salt

To 1 tablespoon of gelatine, softened in ½ cup of cold water add 1 cup of hot tomato juice and pulp. Add seasoned meat. Chill and slice. May be served with salad dressing.

BAKED HASH

1 cup chopped cooked meat

2 cups raw potato, cut fine

1 tablespoon onion juice

2 tablespoons chopped parsley

⅛ teaspoon pepper

¼ cup drippings

½ cup gravy or water

Melt fat in frying pan. Put in all the other ingredients. Cook over a slow fire for ½ hour. Fold and serve as omelet.

MEAT SHORTCAKE

    1½ cups flour

    ½ teaspoon salt

    3 tablespoons shortening

    2 teaspoons baking powder

    2 cups chopped, cooked meat

    1 teaspoon onion juice

    ½ cup gravy or soup stock

    Salt and pepper

    ¾ cup milk and water

[pg 44]

Mix flour, salt and baking powder. Rub in shortening, and mix to dough with milk and water. Roll out to quarter of an inch thickness, bake in layer cake tins. Put together with the chopped meat mixed with the onion and seasoning, and heated hot with the gravy or stock. If stock is used, thicken with a tablespoon of flour mixed with one of butter, or butter substitute. Serve as soon as put together. Cold cooked fish heated in cream sauce may be used for a filling instead of the meat.

SCRAPPLE

Place a pig's head in 4 quarts of cold water and bring slowly to the boil. Skim carefully and season the liquid highly with salt, cayenne and a teaspoon of rubbed sage. Let the liquid simmer gently until the meat falls from the bones. Strain off the liquid, remove the bones, and chop the meat fine.

Measure the liquid and allow 1 cup of sifted cornmeal to 3 cups of liquid. Blend the cornmeal in the liquid and simmer until it is the consistency of thick porridge. Stir in the chopped meat and pour in greased baking pans to cool. One-third buckwheat may be used instead of cornmeal, and any kind of chopped meat can be blended with the pork if desired. Any type of savory herb can also be used, according to taste.

When scrapple is to be eaten, cut into one-half inch slices, dredge with flour, and brown in hot fat.

FISH AS A MEAT SUBSTITUTE

As the main course at a meal, fish may be served accompanied by vegetables or it may be prepared as a "one-meal dish" requiring only bread and butter and a simple dessert to complete a nutritious and well balanced diet. A lack of proper knowledge of selection of fish for the different methods of cooking, and the improper cooking of fish once it is acquired, are responsible to a large extent for the prejudice so frequently to be found against the use of fish.

The kinds of fish obtainable in different markets vary somewhat, but the greatest difficulty for many housekeepers seems to be, to know what fish may best be [pg 45] selected for baking, broiling, etc., and the tests for fish when cooked. An invariable rule for cooking fish is to apply high heat at first, until the flesh is well seared so as to retain the juices; then a lower temperature until the flesh is cooked throughout. Fish is thoroughly cooked when the flesh flakes. For broiling or pan broiling, roll fish in flour or cornmeal, preferably the latter, which has been well seasoned with salt and cayenne. This causes the outside to be crisp and also gives added flavor. Leftover bits of baked or other fish may be combined with white sauce or tomato sauce, or variations of these sauces, and served as creamed fish, or placed in a greased baking dish, crumbs placed on top and browned and served as scalloped fish. Fish canapes, fish cocktail, fish soup or chowder; baked, steamed, broiled or pan broiled fish, entrees without number, and fish salad give opportunity to use it in endless variety.

Combined with starchy foods such as rice, hominy, macaroni, spaghetti or potato, and accompanied by a green vegetable or fruit, the dish becomes a meal. Leftover bits may also be utilized for salad, either alone with cooked or mayonaise salad dressing, or combined with vegetables such as peas, carrots, cucumbers, etc. The addition of a small amount of chopped pickle to fish salad improves its flavor, or a plain or tomato gelatine foundation may be used as a basis for the salad. The appended lists of fish suitable for the various methods of cooking, and the variety in the recipes for the uses of fish, have been arranged to encourage a wider use of this excellent meat substitute, so largely eaten by European epicures, but too seldom included in American menus. During the period of the war, the larger use of fish is a patriotic measure in that it will save the beef, mutton and pork needed for our armies.

[pg 46]

FISH SHORTCAKE

> 2 cups cooked meat or fish
>
> 1 cup gravy or water
>
> 1 teaspoon salt
>
> ⅛ teaspoon cayenne
>
> 1 teaspoon onion juice
>
> 2 cups rye flour
>
> 1 teaspoon of salt
>
> ⅛ teaspoon cayenne
>
> 4 teaspoons baking powder
>
> 4 tablespoons fat
>
> 1 cup gravy, water or milk

Place meat or fish and seasonings in greased dish. Make shortcake by sifting dry ingredients, cut in fat, and add liquid. Place on top of meat or fish mixture. Bake 30 minutes.

CREOLE CODFISH

> 1 cup codfish, soaked over night and cooked until tender
>
> 2 cups cold boiled potatoes
>
> ⅓ cup pimento
>
> 2 cups breadcrumbs

1 cup tomato sauce

Make sauce by melting ¼ cup of fat, adding 2 tablespoons of whole wheat flour.

1 teaspoon salt

⅛ teaspoon pepper

1 teaspoon onion juice, and, gradually

1 cup of tomato and juice

Place the codfish, potatoes and pimento in a baking dish. Cover with the tomato sauce, then the breadcrumbs, to which have been added 2 tablespoons of drippings. Bake brown.

CREAMED SHRIMPS AND PEAS

1 cup shrimps

1 cup peas

2 tablespoons fat

1 teaspoon salt

⅛ teaspoon cayenne

1½ cups milk

2 tablespoons flour

Melt fat, add dry ingredients, and gradually the liquid. Then add fish and peas.

DRESSING FOR BAKED FISH

2 cups breadcrumbs

½ teaspoon salt

⅛ teaspoon pepper (cayenne)

1 teaspoon onion juice

1 tablespoon parsley

1 tablespoon chopped pickle

¼ cup fat

Mix well and fill fish till it is plump with the mixture.
[pg 47]

SHRIMP AND PEA SALAD

1 cup cooked fish

1 cup celery

2 tablespoons pickle

1 cup salad dressing

1 cup peas

FOR DRESSING

1 egg

2 tablespoons flour

1 teaspoon salt

½ teaspoon mustard

2 tablespoons fat

¾ cup milk

¼ cup vinegar

2 tablespoons corn syrup

Directions for making dressing: Mix all ingredients. Cook over hot water until consistency of custard.

FISH CHOWDER

¼ lb. fat salt pork

1 onion

2 cups fish

2 teaspoons salt

⅛ teaspoon pepper

Water to cover

2 cups potatoes, diced

Cook slowly, covered, for ½ hour. Add 1 pint of boiling milk and 1 dozen water crackers.

BAKED FINNAN HADDIE

½ cup each of milk and water, boiling hot

1 fish

Pour over fish. Let stand, warm, 25 minutes. Pour off. Dot with fat and bake 25 minutes. One tablespoon chopped parsley on top.

FISH CROQUETTES

1 cup of cooked fish

1½ cups mashed potato

1 tablespoon parsley

1 egg

½ teaspoon salt

¼ teaspoon cayenne

½ teaspoon celery seed

1 teaspoon lemon juice

Shape as croquette and bake in a moderate oven 25 minutes.

CLAMS A LA BECHAMEL

    1 cup chopped clams

    1½ cups milk

    1 bay leaf

    3 tablespoons fat

    3 tablespoons flour

    ½ teaspoon salt

    ⅛ teaspoon cayenne

    ⅛ teaspoon nutmeg

    1 tablespoon chopped parsley

    1 teaspoon lemon juice

    Yolks of 2 eggs

    ½ cup breadcrumbs

[pg 48]

Scald bay-leaf in milk. Make sauce, by melting fat with flour; add dry ingredients, and gradually add the liquid. Add egg. Add fish. Put in baking dish. Cover top with breadcrumbs. Bake 20 minutes.

SCALLOPED SHRIMPS

    ¼ cup fat

    ¼ cup flour

    ½ teaspoon salt

    ⅛ teaspoon cayenne

    1 cup cooked shrimps

    ½ cup cheese

    ½ cup celery stalk

    1 cup milk

Melt fat, add dry ingredients, and gradually the liquid. Then add fish and cheese. Bring to boiling point and serve.

ESCALLOPED SALMON

    1 large can salmon

    ½ doz. soda crackers

    2 cups thin white sauce

    Salt, pepper

    1 hard-boiled egg

Alternate layers of the salmon and the crumbled crackers in a well-greased baking dish, sprinkling each layer with salt, pepper, the finely chopped hard-boiled egg, and bits of butter or butter substitute, moistening with the white sauce. Finish with a layer of the fish, sprinkling it with the cracker crumbs dotted with butter. Bake in a moderate oven for 30 minutes, or until the top is well browned.

Fish for Frying.—Brook trout, black bass, cod steaks, flounder fillet, perch, pickerel, pompano, smelts, whitefish steak, pike, weakfish, tilefish.

Fish for Boiling.—Cod, fresh herring, weakfish, tilefish, sea bass, pickerel, red snapper, salt and fresh mackerel, haddock, halibut, salmon, sheepshead.

Fish for Baking.—Black bass, bluefish, haddock, halibut, fresh mackerel, sea bass, weakfish, red snapper, fresh salmon, pickerel, shad, muskellunge.

Fish for Broiling.—Bluefish, flounder, fresh mackerel, pompano, salmon steak, black bass, smelts, sea bass steaks, whitefish steaks, trout steaks, shad roe, shad (whole).

[pg 49]

CHEESE AS A MEAT SUBSTITUTE

CHEESE AND BREAD RELISH

>2 cups of stale breadcrumbs
>
>1 cup of American cheese, grated
>
>2 teaspoons of salt
>
>1/8 teaspoon of pepper
>
>2 cups of milk
>
>1 egg
>
>2 tablespoons of fat

Mix well. Bake in a greased dish in moderate oven for 25 minutes.

WELSH RAREBIT

>1 cup of cheese
>
>1 cup of milk
>
>1/4 teaspoon of mustard
>
>1/8 teaspoon of pepper
>
>2 tablespoons of flour
>
>1 teaspoon of fat
>
>1 teaspoon of salt
>
>1 egg

Put milk and cheese in top of double boiler over hot water. Heat until cheese is melted. Mix other ingredients. Add to cheese and milk. Cook five minutes, stirring constantly, and serve at once on toast.

MACARONI WITH CHEESE

Over 1 cup macaroni, boiled in salted water, pour this sauce:

2 tablespoons flour

2 tablespoons fat

1 cupful milk

½ teaspoon salt

⅛ teaspoon pepper

½ cup grated American cheese

Melt fat, add dry ingredients. Add liquid slowly. Bring to boiling point. Add cheese. Stir until melted. Pour over macaroni.

CHEESE AND CABBAGE

2 cups cooked cabbage

¼ cup fat

¼ cup flour

⅛ teaspoon cayenne

1½ cups milk

1 cup grated cheese

1 teaspoon salt

Melt fat, add dry ingredients. Add milk gradually. When at boiling point, add cheese. Pour over cabbage in greased dish and bake 20 minutes. Buttered crumbs may be put on top before baking if desired.

NUT AND CHEESE CROQUETTES

2 cups stale breadcrumbs

1 cup milk

1 yolk of egg

1 cup chopped nuts

⅛ teaspoon salt

⅛ teaspoon cayenne

½ cup grated cheese

Shape and roll in dried breadcrumbs. Bake 20 minutes.
[pg 50]

CHEESE WITH TOMATO AND CORN

1 tablespoon fat

¾ cup cooked corn

½ cup tomato purée

1 teaspoon salt

2 cups grated cheese

¼ cup pimento

1 egg

½ teaspoon paprika

Heat purée. Add fat, corn, salt, paprika and pimento. When hot, add cheese. When melted, add yolk. Cook till thick. Serve on toast.

CHEESE AND CELERY LOAF

½ loaf thinly sliced bread

1 cup cheese

½ teaspoon salt

¼ teaspoon cayenne

¼ cup fat

1 teaspoon Worcestershire sauce

2 eggs

½ cup milk

½ cup cooked celery knob or celery

Mix all ingredients except milk and bread. Spread on bread. Pile in baking dish. Pour milk over the mixture. Bake in a moderate oven until firm in center. Serve hot.

FARINA AND CHEESE ENTREE

    1 cup cooked farina or rice

    1 cup cheese

    1 cup nuts

    1 cup milk

    1 egg

    1 teaspoon salt

    1/8 teaspoon cayenne

Mix all thoroughly. Bake in greased dish 30 minutes.

BOSTON ROAST

    1 teaspoon onion juice

    1 cup grated cheese

    1 teaspoon salt

    1/8 teaspoon cayenne

    1 cup beans (kidney)

    About 1 cup breadcrumbs

Soak and cook beans. Mix all ingredients into loaf. Baste with fat and water. Bake 30 minutes. Serve with tomato sauce.

SPINACH LOAF

    1 cup spinach

    1 cup cheese

    1/8 teaspoon cayenne

½ cup breadcrumbs

1 tablespoon fat

¼ teaspoon salt

Mix and bake in greased dish 20 minutes.

[pg 51]

CHEESE FONDUE

1 cup breadcrumbs

1 cup milk

1 cup cheese

1 egg

2 tablespoons fat

⅛ teaspoon salt

Soak bread 10 minutes in milk. Add fat and cheese. When melted, add egg and seasoning. Cook in double boiler or bake 20 minutes.

RICE-CHEESE RAREBIT

¼ cup fat

¼ cup flour

1 teaspoon salt

1½ cups tomato juice and pulp

1 cup cheese

1 cup cooked rice

⅛ teaspoon cayenne

Melt fat. Add dry ingredients. Add liquid slowly. When at boiling point, add cheese and rice. Serve hot.

POLENTA

    1 cup cooked cornmeal mush

    ½ teaspoon salt

    ½ cup cheese

    ⅛ teaspoon pepper

While mush is hot place ingredients in layers in baking dish. Bake 20 minutes.

CHEESE SAUCE

    ¼ cup fat

    ½ cup flour

    1 teaspoon salt

    2 cups milk

    ½ cup cheese

    ¼ teaspoon cayenne

Prepare same as tomato sauce. Serve with rice or spaghetti.

TOMATO CHEESE SAUCE

    1 pt. milk

    ½ teaspoon soda

    ⅔ cup flour

    2 tablespoons fat

    1 pt. tomatoes

    1 cup cheese

For both the sauces, melt fat, add dry ingredients and, gradually, the liquid. When at boiling point, add cheese and serve. This is an excellent sauce for fish.

[pg 52]

CHEESE SAUCE ON TOAST

¼ cup fat

½ teaspoon salt

1 pint milk

¼ cup flour

¼ teaspoon cayenne

1 cup cheese

Make as white sauce and add cheese. Pour over bread, sliced and toasted. Bake in moderate oven.

CHEESE MOLD

½ pint cottage cheese

¼ cup green peppers, chopped

½ cup condensed milk

⅛ teaspoon of cayenne

1 tablespoon of gelatine

2 tablespoons of cold water

1 teaspoon salt

Soak the gelatine in the cold water until soft. Dissolve over hot water. Add the other ingredients. Chill. Serve as a salad or as a lunch or supper entrée.

CHEESE SOUP

1 quart milk or part stock

¼ cup flour

1 teaspoon salt

¼ cup fat

1 cup cheese

¼ tablespoon paprika

Cream fat and flour; add gradually the liquid, and season. When creamy and ready to serve, stir in the cheese, grated.

CHEESE BISCUIT

1 cup flour

¼ teaspoon salt

½ cup water

3 teaspoons baking powder

1 tablespoon butter or fat

8 tablespoons grated cheese

Mix like drop baking powder biscuit. Bake 12 minutes in hot oven. This recipe makes twelve biscuits. They are excellent to serve with a vegetable salad as they are high in nutrition.

CELERY-CHEESE SCALLOP

1½ cups breadcrumbs

2 cups milk

3 cups chopped celery

1 cup shaved cheese

Cook celery till tender. Put layer of crumbs in greased baking dish, then celery; cover with cheese and sprinkle with salt and pepper. Repeat to fill dish. Turn in boiling hot milk with 1 cup of celery water. Bake for 30 minutes.

[pg 53]

MEAT SUBSTITUTE DISHES

CORN AND OYSTER FRITTERS

    1 cup flour

    2 teaspoons baking powder

    ½ teaspoon salt

    ¼ teaspoon pepper

    ¼ cup milk

    1 egg

    6 oysters

    2 full tablespoons Kornlet

Sift dry ingredients, add milk, egg and Kornlet. Add oysters last. Fry in deep fat, using a tablespoonful to an oyster.

SALMON LOAF

    2 cups cooked salmon

    1 cup grated breadcrumbs

    2 beaten eggs

    ½ cup milk

    ½ teaspoon paprika

    ½ teaspoon salt

    1 tablespoon chopped parsley

    1 teaspoonful onion juice

Mix thoroughly. Bake in greased dish 30 minutes.

BAKED LENTILS

Two cups lentils that have been soaked over night. Boil until soft, with 2 small onions and 1 teaspoon each of thyme, savory, marjoram, and 4 cloves. Drain. Add 1 teaspoon of salt, and put into baking dish. Dot with fat. Bake for 30 minutes.

HOMINY CROQUETTES

- 1 cup of cooked hominy
- ½ cup nuts
- 1 tablespoon corn syrup
- 1 teaspoon of salt
- ⅛ teaspoon of pepper
- 1 egg
- 1 tablespoon melted fat

Mix and roll in dried breadcrumbs and bake in oven 20 minutes.

MEATLESS SAUSAGE

- 1 cup soaked and cooked dried peas, beans, lentils or lima beans
- ½ cup dried breadcrumbs
- ¼ cup fat
- 1 egg
- ½ teaspoon salt
- 1 teaspoon sage

Mix and shape as sausage. Roll in flour and fry in dripping.

[pg 54]

RICE AND NUT LOAF

    1 cup boiled rice or potato

    1 cup peanuts

    ⅔ cup dried breadcrumbs

    ¾ cup milk

    2 teaspoons salt

    ⅛ teaspoon pepper

    ⅛ teaspoon cayenne

    2 tablespoons fat

Mix well. Bake in greased pan 30 minutes.

SOY BEAN CROQUETTES

    2 cups baked or boiled soy beans

    1½ tablespoons molasses

    2 tablespoons butter or drippings

    1 teaspoon salt

    1 tablespoon vinegar

    Pepper to taste

    1 egg

    1 scant cup breadcrumbs

When the beans are placed on to boil, put tablespoon fat and half an onion with them. After draining well, put through the food-chopper, keeping the liquid for soup stock. Mix all the ingredients, beating the egg white before adding. Form into balls or cylinders, dip in the leftover egg yolk, to which a few drops of water have been added, and then coat with stale bread or cracker crumbs. Be sure the croquettes are well covered, then fry brown. Serve with cream sauce or with scalloped or stewed tomatoes. With a green salad, this is a complete meal.

LEGUME LOAF

⅓ cup dried breadcrumbs

2 tablespoons corn syrup

1 egg

1 teaspoon salt

⅛ teaspoon pepper

2 teaspoons chopped nuts

1 teaspoon onion juice

3 tablespoons fat

¾ cup milk

½ cup pulp from peas, beans or lentils, soaked and cooked until tender

Mix well. Bake in greased pan 30 minutes. Serve with tomato sauce, or white sauce, with 2 tablespoons nuts, or 2 teaspoons horseradish added.

VEGETABLE LOAF

One cup peas, beans or lentils soaked over night, then cooked until tender. Put through colander. To 2 cups of mixture, add:

2 eggs

¾ cup dried breadcrumbs

2 teaspoons poultry seasoning

2 teaspoons celery salt

½ cup whole wheat flour

1½ cups tomato juice and pulp

2 teaspoons onion juice

½ teaspoon salt

2 cups chopped peanuts

Mix thoroughly. Place in greased baking dish. Bake 30 minutes.
[pg 55]

KIDNEY BEAN SCALLOP

Two cups kidney beans, soaked over night. Cook until tender. Drain.

To each 2 cups of beans, add:

    2 tablespoons fat

    1 tablespoon chopped onion

    ¼ cup tomato pulp

    1 teaspoon salt

    ⅛ teaspoon pepper

Mix thoroughly. Place in greased baking dish. Cover with 2 cups crumbs, to which have been added 2 tablespoons melted fat. Bake 30 minutes in moderate oven.

VENETIAN SPAGHETTI

    1 cup cooked spaghetti or macaroni

    1 cup carrots

    1 cup turnips

    1 cup cabbage

    2 cups milk

    ½ cup onions

    ¼ cup fat

    ¼ cup flour

1 teaspoon salt

½ cup chopped peanuts

Pepper

Cook spaghetti until tender (about 30 minutes). Cook vegetables until tender in 1 quart water, with 1 teaspoon of salt added. Melt fat, add dry ingredients, add milk gradually and bring to boiling point each time before adding more milk. When all of milk is added, add peanuts. Put in greased baking dish one-half of spaghetti, on top place one-half of vegetables, then one-half of sauce. Repeat, and place in moderately hot oven 30 minutes.

HORSERADISH SAUCE TO SERVE WITH LEFT-OVER SOUP MEAT

3 tablespoons of horseradish

1 tablespoon vinegar

¼ teaspoon salt

⅛ teaspoon cayenne

½ cup of thick, sour cream, and

1 tablespoon corn syrup, or

4 tablespoons of condensed milk

Mix and chill.

BROWN SAUCE FOR LEFTOVER MEATS

⅓ cup drippings

¼ cup of whole wheat flour

⅛ teaspoon pepper

1½ cups meat stock or water

1 teaspoon salt

Melt the fat and brown the flour in it. Add the salt and pepper and gradually the meat stock or water. If water is used, add 1 teaspoon of kitchen bouquet. This may be used for leftover slices or small pieces of any kind of cooked meat.

[pg 56]

FOOD WILL WIN THE WAR
DON'T WASTE IT

"To provide adequate supplies for the coming year is of absolutely vital importance to the conduct of the war, and without a very conscientious elimination of waste and very strict economy in our food consumption, we cannot hope to fulfill this primary duty."

WOODROW WILSON.

SAVE SUGAR

Reasons Why Our Government Asks Us to Save Sugar With Practical Recipes for Sugarless Desserts, Cakes, Candies and Preserves.

One ounce of sugar less per person, per day, is all our Government asks of us to meet the world sugar shortage. One ounce of sugar equals two scant level tablespoonfuls and represents a saving that every man, woman and child should be able to make. Giving up soft drinks and the frosting on our cakes, the use of sugarless desserts and confections, careful measuring and thorough stirring of that which we place in our cups of tea and coffee, and the use of syrup, molasses or honey on our pancakes and fritters will more than effect this saving.

It seems but a small sacrifice, if sacrifice it can be called, when one recognizes that cutting down sugar [pg 58] consumption will be most beneficial to national health. The United States is the largest consumer of sugar in the world. In 1916 Germany's consumption was 20 lbs. per person per year, Italy's 29 to 30 lbs., that of France 37, of England 40, while the United States averaged 85 lbs. This enormous consumption is due to the fact that we are a nation of candy-eaters. We spend annually $80,000,000 on confections. These are usually eaten between meals, causing digestive disturbances as well as unwarranted expense. Sweets are a food and should be eaten at the close of the meal, and if this custom is established during the war, not only will tons of sugar be available for our Allies, but the health of the nation improved.

The average daily consumption of sugar per person in this country is 5 ounces, and yet nutritional experts agree that not more than 3 ounces a day should be taken. The giving up of one ounce per day will, therefore, be of great value in reducing many prevalent American ailments. Flatulent dyspepsia, rheumatism, diabetes, and stomach acidity are only too frequently traced to an oversupply of sugar in our daily diet.

Most persons apparently think of sugar merely as a sweetening agent, forgetting entirely the fact that it is a most concentrated food.

It belongs to what is called the carbohydrate group, upon which we largely depend for energy and heat. It is especially valuable to the person doing active physical work, the open-air worker, or the healthy, active, growing child, but should be used sparingly by other classes of people. Sugar is not only the most concentrated fuel food in the dietary, but it is one that is very readily utilized in the body, 98 per cent. of it being available for absorption, while within thirty minutes of the time it is taken into the system part of it is available for energy.

As a food it must be supplied, especially to the classes [pg 59] of people mentioned above, but as a confection it can well be curtailed. When it is difficult to obtain, housekeepers must avail themselves of changed recipes and different combinations to supply the necessary three ounces per day and to gain the much-desired sweet taste so necessary to many of our foods of neutral flavor with which sugar is usually combined.

Our grandmothers knew how to prepare many dishes without sugar. In their day lack of transportation facilities, of refining methods and various economic factors made molasses, sorghum, honey, etc., the only common methods of sweetening. But the housekeeper of to-day knows little of sweetening mediums except sugar, and sugar shortage is to her a crucial problem. There are many ways, however, of getting around sugar shortage and many methods of supplying the necessary food value and sweetening.

By the use of marmalades, jams and jellies canned during the season when the sugar supply was less limited, necessity for the use of sugar can be vastly reduced. By the addition to desserts and cereals of dried fruits, raisins, dates, prunes and figs, which contain large amounts of natural sugar, the sugar consumption can be greatly lessened. By utilizing leftover syrup from canned or preserved fruits for sweetening other fruits, and by the use of honey, molasses, maple sugar, maple syrup and corn syrup, large quantities of sugar may be saved. The substitution of sweetened condensed milk for dairy milk in tea, coffee and cocoa—in fact, in all our cooking processes where milk is required—will also immeasurably aid in sugar conservation. The substitutes mentioned are all available in large amounts. Honey is especially valuable for children, as it consists of the more simple sugars which are less irritating than cane sugar, and there is no danger of acid stomach from the amounts generally consumed.

[pg 60]

As desserts are the chief factor in the use of quantities of sugar in our diet, the appended recipes will be of value, as they deal with varied forms of nutritious, attractive sugarless desserts. It is only by the one-ounce savings of each individual member of our great one hundred million population that the world sugar shortage may be met, and it is hoped every housekeeper will study her own time-tested recipes with the view of utilizing as far as possible other forms of sweetening. In most recipes the liquid should be slightly reduced in amount and about one-fifth more of the substitute should be used than the amount of sugar called for.

With a few tests along this line one will be surprised how readily the substitution may be made. If all sweetening agents become scarce, desserts can well be abandoned. Served at the end of a full meal, desserts are excess food except in the diet of children, where they should form a component part of the meal.

[pg 61]

SUGARLESS DESSERTS

CRUMB SPICE PUDDING

    1 cup dry bread crumbs

    1 pint hot milk

    Let stand until milk is absorbed.

    ¼ teaspoon salt

    ½ cup molasses

    ¼ teaspoon cinnamon

    1 egg

    ½ teaspoon mixed spices, cloves, nutmeg, allspice, mace and ginger

    ⅔ cup raisins, dates and prunes (steamed 5 minutes)

Mix and bake 45 minutes.

TAPIOCA FRUIT PUDDING

    ½ cup pearl tapioca or sago

    3 cups water

    ¼ lb. dried apricots, prunes, dates or raisins

    ⅛ teaspoon salt

    2 tablespoons fat

    ½ cup corn syrup

Soak fruit in water 1 hour. Add other ingredients. Cook directly over fire 5 minutes, then over hot water until clear, about 45 minutes.

MARMALADE PUDDING

    6 slices stale bread

¼ cup fat

2 egg yolks

1 tablespoon corn syrup

⅛ teaspoon salt

1 cup milk

1 cup marmalade or preserves

Mix eggs, corn syrup, salt and milk. Dip bread and brown in frying pan. Spread with marmalade or preserves. Pile in baking dish. Cover with any of the custard mixture which is left. Cover with meringue. Bake 15 minutes.

PRUNE ROLL

2 cups whole wheat flour

½ cup milk

1 tablespoon fat

2 tablespoons sugar

⅛ teaspoon salt

1 egg

½ lb. washed and scalded prunes, dates, figs or raisins

2 teaspoons baking powder

To prunes, add ½ cup water and soak 10 minutes. Simmer in same water until tender (about 10 minutes). Drain prunes and mash to a [pg 62] pulp. Mix flour, baking powder and salt. Add beaten egg and milk. Mix to a dough. Roll out thin, spread with prune pulp, sprinkle with two tablespoons sugar. Roll the mixture and place in greased baking dish. Bake 30 to 40 minutes. Take half cup of juice from prunes, add 1 tablespoon corn syrup. Bring to boiling point. Serve as sauce for prune roll.

MARMALADE BLANC MANGE

    1 pint milk

    1/8 cup cornstarch

    2 yolks of eggs

    1/3 cup orange marmalade

    1/2 teaspoon vanilla

    Few grains of salt

Mix cornstarch with 1/4 cup of cold milk. Scald rest of milk, add cornstarch, and stir until thick. Cook over hot water 20 minutes. Add rest of ingredients. Cook, stirring 5 minutes. Chill and serve with two whites of eggs, beaten stiff, to which has been added 2 tablespoons orange marmalade. Two ounces grated chocolate and 1/3 cup corn syrup may be substituted for marmalade.

COFFEE MARSHMALLOW CREAM

    2 cups strong boiling coffee

    2 tablespoons gelatine (granulated)

    2 tablespoons cold water

    1/4 cup corn syrup

    1 cup condensed milk

    1/2 teaspoon vanilla

Soak gelatine in cold water until soft. Add coffee and stir until dissolved. Add other ingredients. Chill. One-quarter cup of marshmallows may be cut up and added just before chilling.

FRUIT PUDDING

    2 cups of left-over canned fruit or cooked dried fruit

    2 cups of the juice or water

    1/4 cup corn syrup

2 tablespoons gelatine

1 tablespoon lemon juice

Soften the gelatine in 2 tablespoons of the juice or water. Add the rest of the fruit after it has been heated. When the gelatine is dissolved, add the fruit, lemon juice and corn syrup. Pour in mold.

CEREAL AND DATE PUDDING

1 cup cooked cereal

2 cups milk

1½ tablespoons fat

1 cup dates

¼ cup corn syrup

½ teaspoon salt

1 teaspoon grated lemon rind

½ teaspoon vanilla

1 egg

[pg 63]

Cook over hot water until thick, and boil or bake 20 minutes. Serve with hot maple syrup.

BAKED APPLES WITHOUT SUGAR

Fill cored apples with 1 tablespoon honey, corn syrup, chopped dates, raisins, marmalade, or chopped popcorn mixed with corn syrup in the proportion of two tablespoons of syrup to a cup of corn. Put one-quarter inch of water in pan. Bake until tender and serve apples in pan with syrup as sauce.

APPLES AND POPCORN

Core apples. Cut just through the skin around the center of the apple. Fill the center with popcorn and 1 teaspoon of corn syrup. Bake 30 minutes.

MAPLE RICE PUDDING

½ cup rice

1½ cups milk

¼ teaspoon cinnamon

⅛ teaspoon salt

⅓ cup maple syrup

½ cup raisins

1 egg

Cook in top of double boiler or in steamer 35 minutes.

ECONOMY PUDDING

1 cup cooked cereal

½ cup corn syrup

¼ teaspoon mapline

½ cup milk

½ cup chopped nuts

½ cup raisins or dates

1 egg

Cook in double boiler until smooth. Serve cold with cream or place in baking dish and bake 20 minutes.

OATMEAL AND PEANUT PUDDING

2 cups cooked oatmeal

1 cup sliced apple

1 cup peanuts

½ cup raisins

⅓ cup molasses

½ teaspoon cinnamon

⅛ teaspoon salt

Mix and bake in greased dish for 30 minutes. Serve hot or cold. This is a very nourishing dish.

[pg 64]

CHOCOLATE BLANC MANGE

1 pint milk

⅓ cup cornstarch

⅓ cup corn syrup

1 egg

1 teaspoon vanilla

⅛ teaspoon salt

2 oz. grated chocolate

Mix cornstarch with ¼ cup cold milk. Scald rest of milk. Add cornstarch. Cook until thick. Add a little of the hot mixture to the chocolate when melted. Mix all ingredients and cook 5 minutes, stirring constantly. Chill and serve with plain or chopped nuts.

OATMEAL FRUIT PUDDING

2 cups cooked oatmeal

⅛ cup molasses

1 cup raisins

⅛ teaspoon salt

½ cup chopped nuts

1 egg (beaten)

Mix well. Bake in greased baking dish 30 minutes

JELLIED PRUNES

½ lb. prunes

2½ cups cold water

2 tablespoons granulated gelatine

½ cup corn syrup or ¼ cup sugar

2 teaspoons grated lemon or orange rind

Soak washed and scalded prunes in 2 cups cold water 10 minutes. Simmer until tender (about 10 minutes). Soak gelatine in ½ cup cold water. When soft, add to hot prune mixture. When gelatine is dissolved, add other ingredients and place in mold. Chill, and stir once or twice while chilling to prevent prunes settling to bottom of mold.

APPLE PORCUPINES

Core 6 apples. Cut line around apple just through skin. Fill center with mixture of one-quarter cup each of dates, nuts and figs or marmalade, to which has been added one-quarter cup corn syrup or honey. Bake 30 minutes with one-quarter inch water in baking pan. Stick outside of apple with blanched almonds to make porcupine quills.

SCALLOPED FRUIT PUDDING

2 tablespoons melted fat

2 cups crumbs

½ cup of fruit juice or water

¼ cup corn syrup

2 cups of left-over canned or cooked dried fruit

Put one-quarter of the crumbs on the bottom of a buttered baking pan. Cover with one-half the fruit, one-half the corn syrup, one-half [pg 65] the liquid, one-quarter of the crumbs; the other half of the fruit, juice and corn syrup, and the rest of the crumbs, on top. Bake 20 minutes in a hot oven.

PRUNE FILLING FOR PIE

    ½ lb. pitted prunes

    ⅓ cup corn syrup, or 2 tablespoons sugar

    1 cup water

    2 teaspoons lemon rind

    ½ tablespoon fat

    1 tablespoon cornstarch

Wash and scald prunes. Soak ten minutes in the water. Simmer until tender. Rub through colander. Add other ingredients, well blended. Bring to boiling point. Use as filling for pastry.

APPLE AND DATE FILLING

    2 cups apples

    1 cup dates

    1 tablespoon, fat

    1 teaspoon lemon rind

    ¼ cup water

Mix all and use as filling for double crust, or cook until apples are tender. Mix well and use as filling for tarts, etc.

LEMON FILLING FOR PIE

    1½ cups corn syrup

    1½ cups water

⅓ cup cornstarch

2 eggs

1 tablespoon lemon rind

½ cup lemon juice (2 lemons)

⅛ teaspoon salt

Mix cornstarch and 1 cup water. Add to corn syrup. Cook over direct flame until thick. Cook over hot water 20 minutes. Mix other ingredients. Add one-half cup water and add to other mixture. Cook 5 minutes and use as filling—hot or cold.

SOUR CREAM FILLING FOR CAKE

1 cup sour cream (heated)

1 cup chopped nuts

2 tablespoons corn syrup

1 teaspoon gelatine

2 tablespoons cold water

Soften gelatine in cold water. Add heated cream and when dissolved add other ingredients. Chill and use for cake filling. This is a good way of using up leftover cream which has turned.

[pg 66]

MOCK MINCE MEAT FILLING FOR PIE

1 cup cranberries, chopped

1 cup raisins

1 cup corn syrup

2 tablespoons flour mixed with ¼ cup cold water

2 tablespoons fat

Mix all. Bring to boiling point and place in double crust pastry or cook until thick and use as filling for tarts.

PUMPKIN FILLING FOR PIE

    2 cups stewed pumpkin

    1 cup corn syrup

    1 egg

    2 tablespoons flour

    1 teaspoon cinnamon

    ¾ teaspoon nutmeg

    ¼ teaspoon allspice

    ⅛ teaspoon ginger

    1 teaspoon vanilla

    ⅛ teaspoon salt

    1½ cups milk

Mix all ingredients and bake in double crust pastry, or cook and serve in cooked single crust with meringue.

MERINGUE FOR CHOCOLATE, LEMON OR PUMPKIN PIE

    2 egg whites

    2 tablespoons corn syrup

Beat whites until very stiff. Add corn syrup by folding in. Do not beat.

WHEATLESS, EGGLESS, BUTTERLESS, MILKLESS, SUGARLESS CAKE

    1 cup corn syrup

    2 cups water

2 cups raisins

2 tablespoons fat

1 teaspoon salt

2 teaspoons cinnamon

1 teaspoon nutmeg

1½ cups fine cornmeal, 2 cups rye flour; or, 3½ cups whole wheat flour

1½ teaspoons baking powder, or, ½ teaspoon soda

Cook corn syrup, water, raisins, fat, salt and spices slowly 15 minutes. When cool, add flour, soda or baking powder, thoroughly blended. Bake in slow oven 1 hour. The longer this cake is kept, the better the texture and flavor. This recipe is sufficient to fill one medium-sized bread pan.

SOUR MILK GINGER BREAD

2 tablespoons fat

¼ cup molasses

1 egg

½ teaspoon salt

½ cup sour milk

1 teaspoon soda

2 cups whole wheat flour

1 teaspoon ginger

[pg 67]

Mix soda and molasses. Add other ingredients. Bake in muffin pans 20 minutes or loaf 40 minutes.

MAPLE CAKE

    ¼ cup fat

    1 cup corn syrup

    1½ teaspoons mapline

    1 egg

    1 teaspoon baking powder

    1¼ cups whole wheat flour

    ¼ teaspoon soda

    ¼ cup milk

    ½ teaspoon vanilla

    ½ cup coarsely cut nuts

Cream fat, syrup and mapline. Add beaten egg. Sift dry ingredients and add alternately with milk. Add flavoring and nuts last. Beat well. Bake 20 minutes in layer pan. This quantity makes one layer.

COCOANUT SURPRISE

    6 slices of bread cut in half

    ½ cup of milk

    1 egg yolk

    1 tablespoon corn syrup

    2 tablespoons cocoanut

    Tart jelly

Mix milk, egg yolk and corn syrup. Dip bread in this mixture and brown in frying pan, with small amount of fat. Spread with currant or other tart jelly, preserve or marmalade. Sprinkle with cocoanut and serve as cakes.

SOY BEAN WAFERS

    1 cup soy beans, finely chopped

    ½ cup butter or shortening

    ¼ cup sugar

    ⅓ cup corn syrup

    ½ teaspoon lemon or vanilla

    ½ cup flour

    1 egg

    2 teaspoons baking powder

Soak beans over night, boil for 1 hour. Drain. Cool and put through food-chopper. Cream butter and sugar, add beans, egg. Sift flour with baking powder and add to first mixture. Drop by teaspoonfuls on a baking sheet and bake 8 minutes in a hot oven.

APPLE SPICE CAKE

    ½ cup fat

    ½ cup sugar

    1 beaten egg

    ⅓ cup molasses

    ½ cup tart apple sauce

    ½ cup raisins, dates, prunes or currants (chopped)

    1½ cups flour

    ½ teaspoon allspice

    ¼ teaspoon cloves

    ½ teaspoon nutmeg

[pg 68]

Cream fat and sugar. Add egg. Alternate dry ingredients (which have been sifted together) with the liquid. Add fruit last. Beat well. Bake as loaf about 15 minutes, or in muffin pans about 25 minutes.

CRISP GINGER COOKIES

1 cup of molasses

2 tablespoons of fat

1 teaspoon soda and 1 teaspoon water (hot)

1 cup of flour

1 tablespoon ginger

½ teaspoon cloves

½ teaspoon cinnamon

½ teaspoon salt

About 3 cups flour

Heat molasses and fat until fat is melted. Sift spices with one cup of flour. Dissolve soda in one teaspoon of hot water. Combine all and add enough more flour to make dough stiff enough to roll out. Bake 12 to 15 minutes in moderate oven.

SOFT CINNAMON COOKIES

1 cup molasses

2 tablespoons fat

½ cup boiling water

1 cup flour

1 teaspoon soda

½ teaspoon ginger

2 tablespoons cinnamon

⅛ teaspoon salt

½ teaspoon of cloves

Mix molasses, fat, and boiling water. Sift dry ingredients. Add the liquid. Add enough more flour (about four cups) to make dough stiff enough to roll out. Cut and bake about 15 minutes in moderately hot oven.

WARTIME FRUIT CAKE

    1 cup honey or corn syrup

    1 tablespoon fat

    1 egg

    2 cups flour

    1 teaspoon cinnamon

    1 teaspoon cloves

    ⅛ teaspoon salt

    1 cup chopped dates, figs, prunes or raisins

    ¾ teaspoon soda

    ⅔ cup milk

Cream fat, honey and egg. Sift dry ingredients. Add alternately with milk. Bake in loaf 45 minutes in moderate oven.

HOT WATER GINGER CAKES

    1½ cup molasses

    ¾ cup boiling water

    2½ cups flour

    1⅛ teaspoons soda

    1½ teaspoons ginger

    ¾ teaspoon salt

    ¼ cup fat

[pg 69]

Sift dry ingredients. Mix fat, molasses and boiling water. Add dry ingredients. Beat briskly for a few minutes, and pour into greased muffin pans. Bake twenty to thirty minutes in moderate oven.

SPICED OATMEAL FRUIT CAKES

 1¾ cups whole wheat flour

 ¾ cup cooked oatmeal

 ⅔ cup corn syrup

 ½ cup raisins, dates, prunes or figs

 ¼ teaspoon soda

 ½ teaspoon baking powder

 1 teaspoon cinnamon

 3 tablespoons fat

Heat the corn syrup and fat. Sift dry ingredients and add to first mixture. Add fruit last. Bake in muffin pans for 30 minutes.

FRUIT WONDER CAKES

 1 doz. salted wafers

 ⅓ cup chopped dates

 ⅓ cup chopped nuts

 1 egg white

 2 tablespoons corn syrup

 ½ teaspoon vanilla

Beat egg white until very stiff. Add other ingredients and place on the wafers. Place under broiler until a delicate brown.

SUGARLESS CANDIES

FRUIT PASTE

 2 teaspoons gelatine

 2 tablespoons cold water

 ⅓ cup corn syrup

 2 teaspoons cornstarch

 ¼ cup chopped nuts

 ½ cup chopped dates

 ½ cup chopped raisins

 ¼ teaspoon vanilla

Mix cornstarch with 1 tablespoon cold water. Heat corn syrup to the boil, add cornstarch and cook for three minutes. Soften the gelatine in two tablespoons cold water for five minutes; stir into the hot syrup after taking from fire. When gelatine has dissolved add the fruit and nuts and flavoring. Chill, cut in squares, and roll each in powdered sugar.

WARTIME TAFFY

 2 cups corn syrup

 ½ teaspoon soda

 1 teaspoon water

 2 tablespoons vinegar

Boil the syrup for fifteen minutes, then add the soda. Cook until a little snaps brittle when dropped in cold water. Add the vinegar [pg 70] when this stage is reached and pour into oiled pans. When cool enough to handle, pull until white; make into inch-thick rolls and clip off into neat mouthfuls with oiled scissors, or chill and break into irregular pieces when cold.

PEANUT BRITTLE

 1 cup corn syrup

1 tablespoon fat

1 cup peanuts

Boil syrup and fat until brittle when tested in cold water. Grease a pan, sprinkle the roasted and shelled peanuts in it, making an even distribution, then turn in the syrup. When almost cold mark into squares. Cocoanut, puffed wheat or puffed rice may be used for candy instead of peanuts.

RAISIN AND PEANUT LOAF

Put equal quantity of seeded raisins and roasted peanuts through the food chopper, using the coarsest blade. Moisten with molasses just enough so that the mixture can be molded into a loaf. Chill, cut and serve as candy. Chopped English walnuts combined with chopped dates or figs make a very delicious loaf sweetmeat.

POPCORN BALLS AND FRITTERS

1 cup corn syrup

2 tablespoons vinegar

Popcorn

Cook syrup for fifteen minutes, add vinegar, then when a little snaps when dropped in cold water turn over popped corn, mix well, and form into balls with oiled hands, or if fritters are desired, roll out the mass while warm and cut out with a greased cutter.

COCOANUT LOAF

1 cup shredded cocoanut

½ cup chopped dates

¼ cup corn syrup

⅛ teaspoon mapline

Mix corn syrup and mapline. Add enough to the dates and cocoanut to form a stiff cake. Mold into neat square at least an inch thick. Let stand in the refrigerator for one hour, then cut in squares and roll each in cornstarch.

STUFFED DATES

Mix one-half cup each of chopped peanuts and raisins. Add a teaspoon of lemon juice and two tablespoons of cream cheese. Remove stones from fine large dates, and in their place insert a small roll of the cheese mixture. These are nice in place of candy or can be served with salad.

[pg 71]

FRUIT LOAF

> ½ cup raisins
>
> ½ cup nuts
>
> 2 tablespoons honey, maple syrup or corn syrup
>
> ½ cup figs or dates

Put fruit and nuts through the food chopper, using the coarsest blade. Add enough syrup or honey to make a stiff loaf. Place in the refrigerator for one hour; slice and serve in place of candy, rolling each slice in cornstarch.

STUFFED FIGS

Cut a slit in the side of dried figs, take out some of the pulp with the tip of a teaspoon. Mix with one-quarter cup of the pulp and one-quarter cup of finely chopped crystalized ginger, a teaspoon of grated orange or lemon rind; and a tablespoon of lemon juice. Fill the figs with mixture, stuffing them so that they look plump.

SUGARLESS PRESERVES

QUINCE OR PEAR PRESERVES

1 lb. fruit

1 cup corn syrup

¼ lb. ginger root or 2 oz. crystalized ginger

Steam or cook sliced and pared fruit in small amount of water until tender. Add ginger and corn syrup. Cook 20 minutes slowly. Lemon skins may be used instead of ginger root.

APPLE, QUINCE, PEACH, PEAR OR PLUM JAM

1 cup left-over cooked fruit or pulp from skins and core

¾ cup corn syrup

2 tablespoons vinegar

½ teaspoon mixed ground spices, allspice, cloves and nutmeg

Cook slowly until thick.

PUMPKIN OR CARROT MARMALADE

Reduce 1 pint grape juice one-half by boiling slowly. Add 1 cup vegetables (pumpkin or carrot). Add 2 teaspoons spices and 1 cup corn syrup. Boil until of consistency of honey and place in sterilized jars or glasses.

[pg 72]

GRAPE JUICE

5 lb. grapes

1 pint water

1 cup corn syrup

Cook grapes in water until soft. Mash; drain through jelly bag or wet cheesecloth. Add corn syrup. Boil 5 minutes. Put into sterilized bottles. If cork stoppers are used cover them with melted sealing wax.

SYRUP FOR SPICED APPLES, PEARS, PEACHES, GRAPES

>1 cup corn syrup
>
>2 oz. stick cinnamon
>
>12 allspice berries
>
>6 whole cloves
>
>¼ cup vinegar

Boil 5 minutes. Add any fruit and cook slowly 20 minutes or until fruit is clear and syrup thick. If hard fruits, such as pears, quinces, etc., are used, steam for 20 minutes before adding to syrup.

SYRUP FOR CANNED FRUIT

>1 cup corn syrup
>
>1 cup water

Bring to boiling point. Use same as sugar and water syrup.

SYRUP FOR PRESERVED FRUIT

>2 cups crystal corn syrup For each three pounds of fruit
>
>½ cup water

Use same as water and sugar syrup.

CRANBERRY JELLY

>1 pint cranberries
>
>½ cup water
>
>About 1 cup corn syrup

Cook cranberries in water very slowly until tender. Leave whole or press through colander. Measure amount of mixture and add equal amount of corn syrup. Cook slowly until mixture forms jelly when tested on cold plate. Turn into mold which has been rinsed in cold water.

APRICOT AND RAISIN MARMALADE

    1 cup of apricots

    1½ cups cold water

    1 cup corn syrup

    ½ cup chopped seeded raisins

    1 teaspoon orange rind

Soak apricots and raisins in the water two hours. Cook slowly until very soft. Add other ingredients and cook slowly (about 30 minutes) until slightly thick. Place in sterile jars or glasses and seal.

[pg 73]

SAVE FAT

Reasons Why Our Government Asks Us to Save Fat, With Practical Recipes for Fat Conservation

With the world-wide decrease of animal production, animal fats are now growing so scarce that the world is being scoured for new sources of supply. Our Government has asked the housewife to conserve all the fats that come to her home and utilize them to the best advantage. To this end it is necessary to have some knowledge of the character of different fats and the purposes to which they are best adapted.

The word fat usually brings to one's mind an unappetizing chunk of meat fat which most persons cannot and will not eat, and fatty foods have been popularly supposed to be "bad for us" and "hard to digest." Fats are, however, an important food absolutely essential to complete nutrition, which repay us better for the labor [pg 74] of digestion than any other food. If they are indigestible, it is usually due to improper cooking or improper use; if they are expensive, it is merely because they are extravagantly handled. The chief function of fatty food is to repair and renew the fatty tissues, to yield energy and to maintain the body heat. The presence of fat in food promotes the flow of the pancreatic juice and bile, which help in the assimilation of other foods and assist the excretory functions of the intestine. These are badly performed if bile and other digestive fluids are not secreted in sufficient quantity. The absence of fat in the diet leads to a state of malnutrition, predisposing to tuberculosis, especially in children and young persons.

It is claimed that the most serious food shortage in Germany is fat; that the civilian population is dying in large numbers because of the lack of it, and that Von Hindenburg's men will lose out on the basis of fat, rather than on the basis of munitions or military organization. Worst of all is the effect of fat shortage on the children of the nation. Leaders of thought all over Europe assert that even if Germany wins, Germany has lost, because it has sapped the strength of its coming generation.

The term fat is used to designate all products of fatty composition and includes liquid fats such as oils, soft fats such as butter, and hard fats such as tallow. While all fats have practically the same energy-value, they differ widely from each other in their melting point, and the difference in digestibility seems to correspond to the difference in melting point. Butter burns at 240 degrees Fahrenheit, while vegetable oils can be heated as high as 600 degrees Fahrenheit, furnishing a very high temperature for cooking purposes before they begin to burn. The scorching of fat not only wastes the product, but renders it indigestible, even dangerous to some people, and for this reason butter should never be used for frying, as [pg 75] frying temperature is usually higher than 240 degrees. It is well to choose for cooking only those fats which have the highest heat-resisting qualities because they do not burn so easily.

Beginning with the lowest burning point, fats include genuine butter, substitute butters, lard and its substitutes, and end with tallow and vegetable oils. Of the latter, there is a varied selection from the expensive olive oil to the cheaper cottonseed, peanut, cocoanut and corn oils and their compounds and the hydrogenated oils.

The economy of fat, therefore, depends on the choice of the fat used for the various cooking processes as well as the conservation of all fatty residue, such as crackling, leftover frying fats and soup fat. For cooking processes, such as sauteing (pan frying), or deep fat frying, it is best to use the vegetable and nut oils. These are more plentiful, and hence cheaper than the animal fats; the latter, however, can be produced in the home from the fats of meats and leftover pan fats, which should not be overlooked as frying mediums. Butter and butter substitutes are best kept for table use and for flavoring. The hydrogenated oils, home-rendered fats, lard and beef and mutton suet can be used for shortening fats.

In the purchase of meats, the careful housewife should see that the butcher gives her all the fat she pays for, as all fats can be rendered very easily at home and can be used for cooking purposes. Butchers usually leave as large a proportion of fat as possible on all cuts of meat which, when paid for at meat prices, are quite an expensive item. All good clear fat should, therefore, be carefully trimmed from meats before cooking. Few people either like or find digestible greasy, fat meats, and the fat paid for at meat prices, which could have been rendered and used for cooking, is wasted when sent to table.

[pg 76]

There are various methods of conserving fat. First, the economical use of table fats; second, the saving of cooking; and third, the proper use of all types of fat.

Economy in the use of table fats may best be secured by careful serving. One serving of butter is a little thing—there are about sixty-four of them in a pound. In many households the butter left on the plates probably would equal a serving or one-fourth of an ounce, daily, which is usually scraped into the garbage pail or washed off in the dishpan. But if everyone of our 20,000,000 households should waste one-fourth of an ounce of butter daily, it would mean 312,500 pounds a day, or 114,062,500 pounds a year. To make this butter would take 265,261,560 gallons of milk, or the product of over a half-million cows, an item in national economy which should not be overlooked.

When butter is used to flavor cooked vegetables, it is more economical to add it just before they are served rather than while they are cooking. The flavor thus imparted is more pronounced, and, moreover, if the butter is added before cooking, much of it will be lost in the water unless the latter is served with the vegetables. Butter substitutes, such as oleomargarine and nut margarine, should be more largely used for the table, especially for adults. Conserve butter for children, as animal fats contain vitamines necessary for growing tissues. Butter substitutes are as digestible and as nourishing as butter, and have a higher melting point. They keep better and cost less.

Oleomargarine, which has been in existence for fifty years, was first offered to the world in 1870 by a famous French chemist, Mege-Mouries, who was in search of a butter substitute cheap enough to supply the masses with the much-needed food element. He had noticed that the children of the poor families were afflicted with rickets [pg 77] and other diseases which could be remedied by the administration of the right amount of fat. He combined fresh suet and milk and called the product "oleomargarine." In the United States this product is now made of oleo oil or soft beef fat, neutral lard, cottonseed and other oils, churned with a small quantity of milk, and in the finer grades, cream is sometimes used. A certain proportion of butter is usually added, and the whole worked up with salt as in ordinary butter-making.

Owing to the fears of the butter-makers that oleomargarine would supplant their product in popular favor, legislation was enacted that restricted the manufacture of oleo and established a rigid system of governmental inspection, so that the product is now manufactured under the most sanitary conditions which furnishes a cleaner and more reliable product than natural butter.

Nut margarine is a compound of cocoa oil, which so closely resembles butter that only an expert can distinguish it from the natural product. Both these butter substitutes are used in large amounts by the best bakers, confectioners and biscuit manufacturers, and foolish prejudice against butter substitutes should not deter their use in the home.

A large saving in cooking fats can be made by the careful utilization of all fats that come into the home. Beef and mutton suet can be rendered and made available. Fats which have been saved after meals are cooked should be clarified—that is, freed from all objectionable odors, tastes or color—so as to be made available as shortening and frying fats.

The following recipes and suggestions make possible the use of all fats, and as fat shortage is one of the most serious of the world's food problems, it is essential that every housekeeper have a larger knowledge of the utilization and economy of this essential food.

[pg 78]

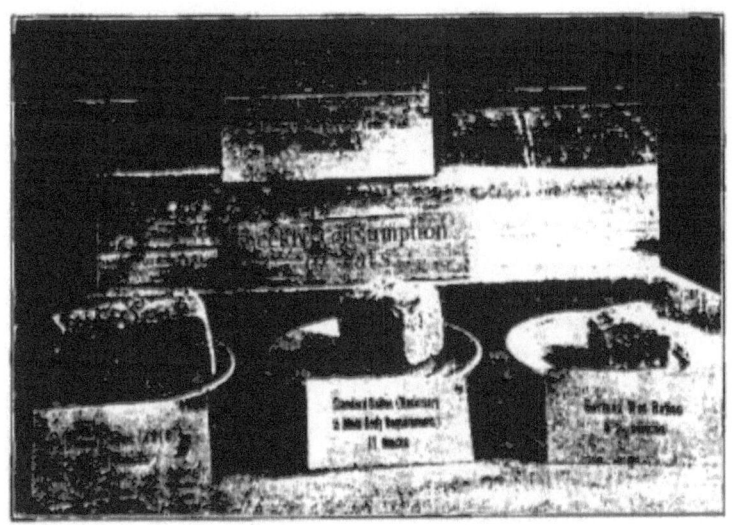

TO RENDER FATS

TO RENDER FAT BY DIRECT METHOD

Run the fat through the household meat grinder or chop fine in the chopping bowl. Then heat in the double boiler until completely melted, finally straining through a rather thick cloth or two thicknesses of cheese cloth, wrung out in hot water. By this method there is no danger of scorching. Fats heated at a low temperature also keep better than those melted at higher temperature. After the fat is rendered, it should be slowly reheated to sterilize it and make sure it is free from moisture. The bits of tissue strained out, commonly known as cracklings, may be used for shortening purposes or may be added to cornmeal which is to be used as fried cornmeal mush.

TO RENDER FAT WITH MILK

To two pounds of fat (finely chopped if unrendered) add one-half pint of milk, preferably sour. Heat the mixture in a double boiler until thoroughly melted. Stir well and strain through a thick cloth or two thicknesses of cheese cloth wrung out in hot water. When cold the fat forms a hard, clean layer and any material adhering to the under side of the fat, may be scraped off. Sour milk being coagulated [pg 79] is preferable to sweet milk since the curd remains on the cloth through which the rendered mixture is strained and is thus more easily separated from the rendered fat which has acquired some of the milk flavor and butter fat.

TO RENDER FAT BY COLD WATER METHOD

Cut fat in small pieces. Cover with cold water. Heat slowly. Let cook until bubbling ceases. Press fat during heating so as to obtain all the oil possible. When boiling ceases strain through cheesecloth and let harden. If desired one-half teaspoon salt, one-eighth teaspoon pepper, 1 teaspoon onion and 1 teaspoon poultry seasoning may be added before straining.

TO RENDER STRONG FLAVORED FATS

To mutton, duck or goose fat add equal amount of beef suet or vegetable fat and render same as suet. This may then be used for shortening, or pan broiling for meat or fish dishes, and not have the characteristic taste of the stronger fats.

When rendering strong mutton, duck or goose fats if a small whole onion is added the strong flavor of the fat is reduced. Remove the onion before straining. It may be used in cooking.

TO CLARIFY FAT

Melt the fat in an equal volume of water and heat for a short time at a moderate temperature. Stir occasionally. Cool and remove the layer of fat which forms on the top, scraping off any bits of meat or other material which may adhere to the other side.

Fats which have formed on top of soups, of cooked meats (such as pot roast, stews), salt meats (such as corned beef, ham, etc.), or strong fats, such as from boiled mutton, poultry and game, may be clarified in this way and used alone or combined with other animal or vegetable fats in any savory dish.

CARE OF FAT AFTER BEING USED FOR COOKING

If fat is used for deep fat frying as croquettes, doughnuts, fritters, etc., while fat is still hot, add a few slices raw potato and allow it to stay in the fat until it is cool. Remove potato—strain fat, allow to harden and it is ready to use. The potato absorbs odors from fat.

HOW TO MAKE SAVORY FATS

FAT 1: To 1 pound of unrendered fat (chopped fine) add 1 slice of onion about one-half inch thick and two inches in diameter, 1 bay leaf, 1 teaspoonful salt, and about one-eighth teaspoonful of pepper. Render in a double boiler and strain.

[pg 80]

FAT 2: To 1 pound unrendered fat (chopped fine) add 2 teaspoonfuls of thyme, 1 slice onion, about one-half inch thick and two inches in diameter, one teaspoonful salt and about one-eighth teaspoonful pepper. Render in a double boiler and strain.

FAT 3: To 1 pound unrendered fat (chopped fine) add 1 teaspoonful thyme, 1 teaspoonful marjoram, one-half teaspoonful rubbed sage, 1 teaspoonful salt, and about one-eighth teaspoonful pepper. Render in a double boiler and strain through fine cloth.

EXTENSION OF TABLE FATS

A. Butter or other fat may be extended to double its original bulk and reduce the cost of the fat 40 per cent. A patented churn, any homemade churn, mayonnaise mixer, or bowl and rotary beater may be used for the purpose. To any quantity of butter heated until slightly soft add equal quantity of milk, place in the churn, add one teaspoon salt for each one pound of butter used. Blend thoroughly in churn, mayonnaise mixer, or in bowl with rotary beater until of even consistency. Place in refrigerator to harden. Vegetable coloring, such as comes with margarine or may be purchased separately, may be added if a deeper yellow color is desired.

B.

    1 lb. butter

    1 quart milk (2 pint bottles preferred)

>1 tablespoon granulated gelatine
>
>1½ teaspoons salt

Soak gelatine in one-half cup of the milk. When softened, dissolve over hot water. Let butter stand in warm place, until soft. Add gelatine mixture, milk and salt and beat with Dover beater until thoroughly mixed (about 15 minutes). Vegetable coloring such as comes with margarine may be added if desired. Do not put on ice.

C.

>1 lb. butter
>
>1 quart milk (2 pint bottles preferred)
>
>1 tablespoon granulated gelatine
>
>1½ teaspoons salt
>
>1 cup peanut butter

Soak gelatine in one-half cup of the milk. When softened, dissolve over hot water. Let butter stand in warm place, until soft. Add gelatine mixture, peanut butter, milk and salt and beat with rotary egg beater until thoroughly blended (about 15 minutes). Vegetable coloring such as comes with margarine may be added if desired. Put in cool place to harden but do not put on ice as the gelatine would cause the mixture to flake. It is preferable to make up this mixture enough for one day at a time only.

D. To 1 pound of butter or butter substitute add one cup peanut butter. Blend thoroughly with wooden spoon or butter paddle; this may be used in place of butter as a new and delightful variation.

[pg 81]

E. To 1 pound softened butter add 1 pound softened butter substitute (oleomargarine, nut margarine, vegetable margarine) or hydrogenated fat. Blend thoroughly with butter paddle or wooden spoon and use as butter.

SUGGESTIONS FOR PASTRY

Whole wheat makes a more tasty crust than bread flour and all rye pastry has even better flavor than wheat flour pastry. Half wheat or rye and the other half cornmeal (white or yellow) makes an excellent pastry for meat or fish pie. If cornmeal is added, use this recipe:

CORNMEAL PASTRY FOR MEAT OR FISH

    ½ cup cornmeal

    ½ cup rye or wheat flour

    2 tablespoons fat

    ⅓ cup cold or ice water

    1 teaspoon baking powder

Sift dry ingredients. Cut in fat. Add water and roll out on well floured board.

PASTRY MADE WITH DRIPPING

Well made, digestible pastry should have a minimum of fat to make a crisp flaky crust. It should be crisp, not brittle; firm, not crumbly. Pastry may be made in large amounts, kept in refrigerator for several days and used as needed. Roll out only enough for one crust at a time as the less pastry is handled, the better.

PLAIN PASTRY

    1 cup flour

    ⅓ cup fat

    ½ teaspoon salt

    About ¼ cup cold or ice water

Mix flour and salt. Cut in fat and add just enough cold or ice water to make the mixture into a stiff dough. Roll out. This recipe makes one crust.

MEAT OR FISH PIE CRUST

>2 cups flour
>
>4 teaspoons baking powder
>
>⅓ cup any kind of dripping
>
>1 cup meat stock or milk
>
>1 teaspoon salt

Sift dry ingredients. Cut in fat if solid, or add if liquid. Stir in meat stock or milk to make a soft dough. Place on top of meat or fish with gravy in greased baking dish and bake 30 to 40 minutes in moderately hot oven.

[pg 82]

VARIOUS USES FOR LEFTOVER FATS

CREOLE RICE

>2 tablespoons savory drippings
>
>3 tablespoons flour
>
>1 teaspoon salt
>
>½ teaspoon cayenne
>
>1½ cups tomato juice and pulp
>
>1 teaspoon onion juice
>
>2 tablespoons chopped green pepper
>
>1 tablespoon chopped olives
>
>1 cup of rice
>
>1 cup water

Wash rice and soak in water 30 minutes. Melt fat, add dry ingredients and gradually the tomatoes. Stir in rice and other ingredients, also the water in which rice was soaked. Cook slowly one-half hour or until rice is tender.

POTATOES ESPAGNOLE

    2 cups pared and sliced potatoes

    2 tablespoons bacon drippings

    2 tablespoons minced onion

    ½ teaspoon salt

    ¼ tablespoon cayenne

    1½ cups boiling water

    1 tablespoon chopped green pepper or pimento

Melt drippings. Add onion and cook until slightly brown. Add other seasonings and water. Pour over potatoes. Let cook slowly in oven until potatoes are tender, about 30 minutes.

DUMPLINGS

    2 cups flour

    1 teaspoon salt

    4 teaspoons baking powder

    2 tablespoons drippings

    1 cup water, meat stock or milk

Sift dry ingredients. Cut in fat. Gradually add liquid to make a soft dough. Roll out, place on greased pan and steam 20 minutes, or drop into stew and cook covered 30 minutes. Serve at once.

POTATO SALAD

    2 cups freshly cooked and diced potatoes

⅓ cup bacon drippings

½ teaspoon salt

2 tablespoons chopped peppers

2 tablespoons vinegar

⅛ teaspoon cayenne

Mix drippings, salt, pepper, vinegar and cayenne. Add to the potatoes and mix thoroughly. Chill and serve. Cold cooked potatoes may be used, but the flavor is better if mixed while potatoes are hot.

SOAP

1 can lye

6 lbs. fat (Fat for soap should be fat which is no longer useful for culinary purposes.)

1 quart cold water

To lye add water — using enamel or agate utensil. When cool add the fat which has been heated until liquid. Stir until of consistency of honey (about 20 minutes). Two tablespoons ammonia or two tablespoons borax may be added for a whiter soap. If stirred thoroughly this soap will float.

[pg 83]

The illustration shows various forms of food waste—the discarded outside leaves of lettuce and cabbage, apple cores and parings, stale bread and drippings.

SAVE FOOD

Reasons Why Our Government Asks Us Not to Waste Food, with Practical Recipes on the Use of Leftovers

Elimination of food waste is to-day a patriotic service. It is also a most effective method of solving our food problem. This country, like all the powers at war, will undoubtedly be called upon to face increasing prices so long as the war continues, and waste in any form is not only needless squandering of the family income, but failure in devotion to a great cause.

Food waste is due to poor selection of raw materials, to careless storage and heedless preparation, to bad cooking, to injudicious serving, and to the overflowing garbage pail.

To select food in such a way as will eliminate waste and at the same time insure the best possible return for money spent, the housekeeper must purchase for nutriment rather than to please her own or the family palate.

When eggs are sixty and seventy cents a dozen their price is out of all proportion to their food value. Tomatoes [pg 84] at five or ten cents apiece in winter do not supply sufficient nutriment to warrant their cost, nor does capon at forty-five cents a pound nourish the body any better than the fricassee fowl at twenty-eight cents. In order to prevent such costly purchasing, a knowledge of food values is necessary. The simplest and easiest way to plan food values is to divide the common food materials into five main groups and see that each of these groups appear in each day's menu.

GROUP 1.—Foods depended on for mineral matters, vegetable acids, and body-regulating substances.

FRUITS

    Apples, pears, etc.,

    Berries,

    Melons,

    Oranges, lemons, all citrus fruits.

VEGETABLES

    Salads, lettuce, celery,

    Potherbs or "greens"

    Tomatoes, squash,

    Green peas, green beans,

    Potatoes and root vegetables.

GROUP 2. — Foods depended on for protein.

    Milk, skim milk, cheese,

    Eggs,

    Meat,

    Poultry,

    Fish,

    Dried peas, beans, cow-peas,

    Nuts.

GROUP 3. — Foods depended on for starch.

    Cereals, grains, meals, flour,

    Cereal breakfast foods,

    Bread,

    Crackers,

    Macaroni and other pastes,

    Cakes, cookies, starchy puddings,

    Potatoes, other starchy vegetables,

    Bananas.

[pg 85]

GROUP 4.—Foods depended on for sugar.

    Sugar,

    Molasses,

    Syrups,

    Dates,

    Raisins,

    Figs.

GROUP 5.—Foods depended on for fat.

    Butter and cream,

    Lard, suet,

    Salt pork and bacon,

    Table and salad oils,

    Vegetable, nut, and commercial cooking fats and oils.

If from each of these groups the housekeeper, when buying, chooses the lowest-cost food, she will provide the necessary nutriment for the least expenditure of money. In war time such marketing is essential.

Other causes of waste in food purchasing may be enumerated as follows: Ordering by telephone. This permits the butcher or grocer, who has no time to make selection of foods, to send what comes ready to hand; whereas if the housekeeper did her own selecting, she could take advantage of special prices or "leaders"—food sold at cost or nearly cost to attract patronage.

Buying out-of-season foods also makes marketing costly. Through lack of knowledge concerning the periods at which certain fruits and vegetables are seasonable, and therefore cheaper and in best flavor, housekeepers frequently pay exorbitant prices for poor flavored, inferior products.

Buying in localities where high rental and neighborhood standards compel the shopkeeper to charge high prices, the consumer pays not only for the rent and the plate glass windows, but for display of out-of-season delicacies, game and luxury-foods. Markets should be selected where food in season is sold; where cleanliness and careful attention prevail rather than showy display.

[pg 86]

Many a dollar is foolishly spent for delicatessen foods. The retail cost of ready prepared foods includes a fraction of the salary of the cook and the fuel, as well as the regular percentage of profit. The food, also, is not so nourishing or flavorsome as if freshly cooked in the home kitchen.

Buying perishable foods in larger quantities than can be used immediately. Too frequently meats, fish, eggs, vegetables, milk and cream are purchased in quantities larger than needed for immediate consumption, and lack of knowledge of use of left-overs causes what is not eaten to be discarded.

Buying non-perishable foods in small quantities instead of in bulk. Food costs on an average 50 to 75 per cent. more when purchased in small quantities. Select a grocer who keeps his goods in sanitary condition and who will sell in bulk; then do your purchasing from him on a large scale and extend the sanitary care to your own storeroom.

Buying foods high in price but low in food value. Asparagus, canned or fresh, is not as nourishing, for instance, as canned corn or beans. Strawberries out of season do not compare with dates, figs or raisins which are to be had at all times.

Buying without planning menus. By this carelessness foods are often purchased which do not combine well, and therefore do not appeal to the appetite, and so are wasted. Unplanned meals also lead to an unconscious extravagance in buying and an unnecessary accumulation of left-overs.

Buying foreign brands when domestic brands are cheaper and often better.

Leaving the trimmings from meats and poultry at the [pg 87] butcher's. Bring these home and fry out the fatty portions for dripping; use all other parts for the stock pot.

Having purchased for nutriment and in sufficiently large quantities to secure bulk rates, careful storage is the next step in the prevention of waste. Flour, cereals and meals become wormy if they are not kept in clean, covered utensils and in a cool place. Milk becomes sour, especially in summer. This can be prevented by scalding it as soon as received, cooling quickly, and storing in a cold place in covered, well-scalded receptacles. Sour milk should not be thrown out. It is good in biscuits, gingerbread, salad dressings, cottage cheese, pancakes or waffles, and bread making.

Meats should not be left in their wrappings. Much juice soaks into the paper, which causes a loss of flavor and nutriment. Store all meat in a cool place and do not let flies come in contact with it.

Bread often molds, especially in warm, moist weather. Trim off moldy spots and heat through. Keep the bread box sweet by scalding and sunning once a week.

Cheese molds. Keep in a cool, dry place. If it becomes too dry for table use, grate for sauces or use in scalloped dishes.

Winter vegetables wilt and dry out. Store in a cool place. If cellar space permits, place in box of sand, sawdust or garden earth.

Potatoes and onions sprout. Cut off the sprouts as soon as they appear and use for soup. Soak, before using, vegetables which have sprouted.

Fruits must be stored carefully so as to keep the skins unbroken. Broken spots in the skin cause rapid decay. Do not permit good fruit to remain in contact with specked or rotted fruit. Stored fruit should be looked over frequently and all specked or rotted fruit removed. [pg 88] Sweet potatoes are an exception. Picking over, aggravates the trouble. See that these vegetables are carefully handled at all times; if rot develops, remove only those that can be reached without danger of bruising the sound roots. Sweet potatoes may also be stored like fruit by spreading over a large surface and separating the tubers so that they do not touch each other.

Berries should be picked over as soon as received and spread on a platter or a large surface to prevent crushing and to allow room for circulation of air.

Lettuce and greens wilt. Wash carefully as soon as received and use the coarse leaves for soup. Shake the water from the crisp portions and store in a paper bag in a cold refrigerator.

Lemons when cut often grow moldy before they are used. When lemons are spoiling, squeeze out the juice, make a syrup of one cup of sugar and one cup of water, boil ten minutes and add lemon juice in any amount up to one cup. Bring to boiling point and bottle for future use. This bottled juice may be used for puddings, beverages, etc. If only a small amount of juice is needed, prick one end of a lemon with a fork. Squeeze out the amount needed and store the lemon in the ice-box.

When we come to waste caused by careless preparation we may be reminded of the miracle of the loaves and fishes—how all the guests were fed and then twelve baskets were gathered up. Often after preparation that which is gathered up to be thrown away is as large in quantity and as high in food value as the portions used.

Vegetables are wasted in preparation by too thick paring, the discarding of coarse leaves such as are found on lettuce, cabbage and cauliflower, discarding wilted parts which can be saved by soaking, throwing away tips and roots of celery and the roots and ends of spinach and dandelions. All these waste products can be cooked [pg 89] tender, rubbed through a sieve and used with stock for vegetable soup, or with skimmed milk for cream soup. Such products are being conserved by the enemy, even to the onion skin, which is ground into bread-making material.

Throwing away the water in which vegetables have been cooked wastes their characteristic and valuable element—the mineral salts. Cooking them so much that they become watery; under-cooking so that they are hard and indigestible; cooking more than is required for a meal; failing to use left-over portions promptly as an entree or for cream soups or scalloped dishes—all these things mean an appalling waste of valuable food material. Good food material is also lost when the water in which rice or macaroni or other starchy food has been boiled is poured down the kitchen sink. Such water should be used for soup making.

Fruits are wasted by throwing away the cores and skins, which can be used for making sauces, jams and jellies, the latter being sweetened with corn syrup instead of sugar.

Rhubarb is wasted by removing the pink skin from young rhubarb, which should be retained to add flavor and color-attractiveness to the dish.

Raw food in quantity is frequently left in the mixing bowl, while by the use of a good flexible knife or spatula every particle can be saved. A large palette knife is as good in the kitchen as in the studio.

The next step in food preparation is cooking, and tons of valuable material are wasted through ignorance of the principles of cooking.

Bad cooking, which means under-cooking, over-cooking or flavorless cooking, renders food inedible, and inedible food contributes to world shortage. [pg 90] Fats are wasted in cooking by being burned and by not being carefully utilized as dripping and shortening. The water in which salt meat, fresh meat, or poultry has been boiled should be allowed to cool and the fat removed before soup is made of it. Such fat can be used, first of all, in cooking, and then any inedible portions can be used in soap making.

Tough odds and ends of meat not sightly enough to appear on the table are often wasted. They can be transformed by long cooking into savory stews, ragouts, croquettes and hashes, whereas, if carelessly and insufficiently cooked, they are unpalatable and indigestible. Scraps of left-over cooked meat should be ground in the food-chopper and made into appetizing meat balls, hashes or sandwich paste. If you happen to have a soft cooked egg left over, boil it hard at once. It can be used for garnishes, sauces, salads or sandwich paste.

Use all bits of bread, that cannot be used as toast, in puddings, croquettes, scalloped dishes or to thicken soup.

Don't throw away cold muffins and fancy breads. Split and toast them for next day's breakfast.

Foods that survive the earlier forms of waste are often lost at table by the serving of portions of like size to all members of the family. The individual food requirements differ according to age, sex, vocation and state of health. Each should be considered before the food is served, then there will be no waste on the plates when the meal is over. The following table, showing the daily requirement of calories for men and women in various lines of work, illustrates this point:

[pg 91]

    WOMEN CALORIES

    Sedentary work ... 2,400

    Active work ... 2,700

    Hard manual labor ... 3,200

    MEN CALORIES

    Sedentary work ... 2,700

    Active work ... 3,450

    Hard manual labor ... 4,150

Although the serving of food should be carefully planned so as to prevent waste, care should be taken that growing children have ample food. It is a mistake to suppose that a growing child can be nourished on less than a sedentary adult. A boy of fourteen who wants to eat more than his father probably needs all that he asks for. We must not save on the children; but it will be well to give them plain food for the most part, which will not tempt them to overeat, and tactfully combat pernickety, overfastidious likes and dislikes.

The United States Food Administration is preaching the gospel of the clean plate, and this can be accomplished by serving smaller portions, insisting that all food accepted be eaten; by keeping down bread waste, cutting the bread at the table a slice at a time as needed; by cooking only sufficient to supply moderately the number to be fed, and no more. It is a false idea of good providing that platters must leave the table with a generous left-over. Waste of cooked food is a serious item in household economy, and no matter how skillfully leftovers are utilized, it is always less expensive and more appetizing to provide fresh-cooked foods at each meal.

One would think that with the various uses to which all kinds of foodstuffs may be put that there would be little left for the yawning garbage pail. But the Secretary of the United States Department of Agriculture is responsible for the statement that $750,000,000 worth of [pg 92] food has been wasted annually in the American kitchen. Undoubtedly a large part of this wastefulness was due to ignorance on the part of the housewife, and the rest of it to the lack of co-operation on the part of the employees who have handled the food but not paid the bills.

According to a well-known domestic scientist, the only things which should find their way to the garbage pail are:

> Egg shells—after being used to clear coffee.
>
> Potato skins—after having been cooked on the potato.
>
> Banana skins—if there are no tan shoes to be cleaned.
>
> Bones—after having been boiled in soup kettle.
>
> Coffee grounds—if there is no garden where they can be used for fertilizer, or if they are not desired as filling for pin-cushions.
>
> Tea leaves—after every tea-serving, if they are not needed for brightening carpets or rugs when swept.
>
> Asparagus ends—after being cooked and drained for soup.
>
> Spinach, etc.—decayed leaves and dirty ends of roots.

If more than this is now thrown away, you are wasting the family income and not fulfilling your part in the great world struggle. Your government says that it is your business to know what food your family needs to be efficient; that you must learn how to make the most of the foods you buy; that it is your duty to learn the nature and uses of various foods and to get the greatest possible nourishment out of every pound of food that comes to your home.

The art of utilizing left-overs is an important factor in this prevention of waste. The thrifty have always known it. The careless have always ignored it. But now as a measure of home economy as well as a patriotic service, the left-over must be handled intelligently.

The following recipes show how to make appetizing dishes from products that heretofore in many homes have found their way to the extravagant pail.

[pg 93]

In these recipes, sauces are prominent because they are of great value in making foods of neutral flavor, especially the starchy winter vegetables, and rice, macaroni and hominy, as attractive as they are nutritious; salads are included, since these serve to combine odds and ends of meats and vegetables; gelatine dishes are provided because gelatine serves as a binder for all kinds of leftovers and is an extremely practical way of making the most rigid saving acceptable; desserts made of crumbs of bread and cake, or left-over cereals, are among the major economies if they are worked out in such a way that they do not involve the extravagant use of other foodstuffs. All the recipes in this economy cook-book have been thoughtfully adapted to the conditions of the time, and will show the practical housekeeper how to supply wholesome, flavorsome food for the least cost.

SAUCES MAKE LEFTOVERS ATTRACTIVE

WHITE SAUCE

¼ cup flour

¼ cup fat

1 teaspoon salt

⅛ teaspoon cayenne

1½ cups milk

Melt fat. Add dry ingredients and a little of the milk. Bring to boiling point. Continue adding milk a little at a time until all is added. Serve with vegetables, fish, eggs, meats.

WHITE SAUCE WITH CHEESE

    ½ cup cheese (cream or American) added to

    1½ cups white sauce

Excellent to serve with macaroni, hominy or vegetables.

WHITE SAUCE WITH SHRIMPS

    ½ cup shrimps

    ¼ teaspoon salt

    1 cup white sauce

Serve on toast, or with starchy vegetables.

[pg 94]

WHITE SAUCE WITH HORSERADISH AND PIMENTO

    ¼ cup horseradish

    1 tablespoon chopped pimento

    1 cup white sauce

Serve with boiled beef, hot or cold, or with cold roast beef.

WHITE SAUCE WITH EGG

    1 cup white sauce

    2 sliced hard-cooked eggs

    ⅛ teaspoon cayenne

    ⅛ teaspoon salt

Excellent for spinach and vegetables, or fish.

BROWN SAUCE

¼ cup fat

⅓ cup flour

1 teaspoon salt

⅛ teaspoon of cayenne

1½ cups brown stock, or

1½ cups water and 2 bouillon cubes

½ teaspoon Worcestershire sauce

Melt fat until brown. Add flour. Heat until brown. Add liquid gradually, letting come to boiling point each time before adding more liquid. When all is added, 1 teaspoon kitchen bouquet may be added if darker color is desired.

BROWN SAUCE WITH OLIVES

1 cup brown sauce

3 tablespoons chopped olives

Make brown sauce as given in foregoing recipe, then while it is hot stir in the chopped olives, and serve.

BROWN SAUCE WITH PEANUTS

1 cup brown sauce

¼ cup chopped peanuts

⅛ teaspoon salt

A good sauce to serve with rice, macaroni, hominy or other starchy foods. It supplies almost a meat flavor to these rather insipid foods.

MUSHROOM SAUCE

    1 cup brown sauce

    ½ cup chopped mushrooms

Add mushrooms to fat and flour before adding liquid. If fresh mushrooms are used, cook for two or three minutes after adding liquid.

[pg 95]

VEGETABLE SAUCES

    ¼ cup fat

    ¼ cup flour

    1 teaspoon salt

    ⅛ teaspoon cayenne

    2 cups vegetable stock,

    or

    1 cup vegetable stock

    1 cup milk.

Vegetable stock is the water in which any vegetable is cooked. Make as white sauce.

DRAWN BUTTER SAUCE

    ⅓ cup butter substitute

    ¼ cup flour

    ½ teaspoon salt

    ⅛ teaspoon cayenne

    1 cup boiling water

    2 tablespoons chopped parsley

Make as white sauce, reserving 2 tablespoons of the fat to add just before serving.

TOMATO SAUCE

¼ cup fat

¼ cup flour

1 teaspoon salt

¼ teaspoon cayenne

1 teaspoon Worcestershire

1 teaspoon onion juice

1½ cups tomato

Melt fat; add dry ingredients and gradually the liquid, letting sauce come to boiling point each time before adding more liquid.

FRUIT SAUCE FOR PUDDING

¼ cup fat

½ cup milk

½ cup powdered sugar

1 teaspoon vanilla, or

1 tablespoon brandy

1 cup mashed cooked fruit

Mix thoroughly. Let chill and serve with steamed or baked pudding.

COCOANUT SAUCE

½ cup milk

½ cup cocoanut and milk

2 tablespoons corn syrup

2 tablespoons cornstarch

1 teaspoon vanilla

Mix ingredients. Bring to boiling point over direct fire. Cook over hot water 20 minutes. Use with leftover stale cake, baked or steamed puddings. If canned cocoanut containing milk is used, plain milk may be omitted.

[pg 96]

MOLASSES SAUCE

1 cup molasses

2 tablespoons fat

1 tablespoon flour, plus

1 tablespoon cold water

1½ tablespoons vinegar

Mix together. Bring to boiling point and serve with any pudding.

FRENCH SAUCE

1 cup (crystal) corn syrup

⅛ teaspoon salt

1 egg

½ cup water

1 tablespoon cream

1 teaspoon vanilla

Beat egg light. Pour on gradually the hot corn syrup and water, beating egg with eggbeater. Add cream and vanilla. Serve at once.

SPICE SAUCE

½ cup corn syrup

1 egg

⅓ cup milk

½ teaspoon cinnamon

½ teaspoon nutmeg

½ teaspoon vanilla

Mix corn syrup and spices. Add beaten yolks and milk. Cook over hot water until thick. Add vanilla and beaten whites. Serve hot or cold.

MAPLE SPICE SAUCE

3 tablespoons fat

⅓ cup maple sugar

2 eggs

½ teaspoon cinnamon

½ teaspoon allspice

½ teaspoon vanilla

⅓ cup milk

Cream fat, sugar and spices. Add beaten yolks and milk. Cook in double boiler until thick. Add vanilla and beaten whites. Serve hot or cold.

TOMATO SAUCE WITH CHEESE

1 cup tomato sauce

½ cup grated cheese

Add cheese while sauce is hot and just before serving. Do not boil sauce after adding cheese.

MEXICAN SAUCE

To one cup tomato sauce, add

    2 tablespoons chopped green pepper

    3 tablespoons chopped celery

    3 tablespoons chopped carrot

[pg 97]

HARD SAUCE

    ⅓ cup butter substitute or hydrogenated oil

    ⅓ cup corn syrup

    ⅓ cup sugar

    1 teaspoon flavoring

Cream all together. This method reduces the necessary sugar two-thirds.

LEMON OR ORANGE SAUCE

    ½ cup corn syrup

    1 tablespoon fat

    ¼ cup lemon juice

    1 teaspoon lemon rind

    2 tablespoons cornstarch

    3 tablespoons lemon juice

    ½ cup orange juice

    2 teaspoons orange rind

    1 tablespoon flour

1 tablespoon water

Mix ingredients. Bring to boiling point and serve.

CRANBERRY SAUCE WITH RAISINS
- 1 cup cranberries
- 1 cup water
- 1 cup corn syrup
- ½ cup raisins or nuts
- 2 tablespoons fat

Cook cranberries in water until they are soft and the water is almost entirely absorbed. Add other ingredients and cook about 20 minutes slowly until thick enough to use as sauce.

THE USE OF GELATINE IN COMBINING LEFTOVERS

LEFTOVER FRUIT MOLD
- 2 tablespoons cold water
- 2 tablespoons gelatine

Let stand until gelatine is soft. Add 1 pint boiling water, or fruit juice from canned fruit.
- ¼ cup lemon juice
- ⅔ cup corn syrup, or
- ½ cup sugar

Stir until gelatine is dissolved. Add 1 cup leftover fruit. Place in mold which has been dipped in cold water. Stir occasionally while hardening so fruit does not settle to the bottom. Or a little gelatine may be poured in mold and allowed to grow almost hard; then some fruit arranged on it and more gelatine poured in. Repeat until mold is filled; then chill, and turn out carefully.

[pg 98]

MOLDED VEGETABLE SALAD

1½ cups boiling tomato juice and pulp

2 tablespoons cold water

2 tablespoons gelatine

1 teaspoon salt

¼ teaspoon paprika

¼ teaspoon Worcestershire sauce

2 cups of any one vegetable, or of mixed vegetables

Soften gelatine in the cold water. Add other ingredients and chill. Stir once or twice while chilling so vegetables do not settle to the bottom.

MOLDED MEAT OR FISH LOAF

2 tablespoons gelatine

2 tablespoons cold water

1 cup boiling gravy, tomato juice, or 1 cup boiling water into which 1 bouillon cube has been dissolved

1 cup left-over meat or fish chopped fine

1 cup chopped celery or cooked vegetable

1 teaspoon salt

⅛ teaspoon cayenne

Soften gelatine in cold water. Add other ingredients. Stir until gelatine is dissolved. Pour into mold dipped into cold water. Chill.

Stir once or twice while hardening so meat does not settle to the bottom. Serve with salad dressing.

RICE IMPERIAL

    1 cup cooked rice

    1 cup corn syrup

    1 tablespoon gelatine

    2 tablespoons water

    ½ cup cherries or other cooked fruit

    ½ cup nuts

    ½ cup juice of fruit

Chill and serve.

CREAM SALAD MOLD

    1 cup cooked salad dressing

    2 tablespoons gelatine

    2 cups any left-over fish, meat or vegetables

    2 tablespoons cold water

Use any well-seasoned salad dressing. Soften the gelatine in the cold water. Dissolve over boiling water. Add to salad dressing. Add other ingredients well seasoned and chill.

CHEESE MOLD

    1 pint cottage cheese

    ½ cup pimento or green pepper

    1 cup milk

    2 teaspoons salt

    ¼ teaspoon cayenne

2 tablespoons granulated gelatine

4 tablespoons cold water

[pg 99]

Soften gelatine in the cold water. Dissolve over hot water. Add all ingredients. Mix thoroughly and place in mold which has been rinsed with cold water. When firm, serve as salad.

FRUIT SPONGE

2 tablespoons gelatine softened in

⅓ cup cold water

1 pint clabbered milk, or fruit juice

1 cup sugar

1 teaspoon vanilla

1 cup crushed fruit

2 egg whites

Mix gelatine with milk. Add sugar. When it begins to thicken, beat with rotary beater. Add vanilla and fruit. Fold in egg whites and turn into mold. Apple sauce, strawberries, rhubarb, pineapple or raspberries may be used.

ORIENTAL SALAD

1 tablespoon gelatine

2 cups boiling water

¾ cup sugar

½ cup lemon juice

½ cup grated cocoanut

2 cups apples, chopped

1 cup celery

½ cup chopped nuts

3 pimentoes

1 tablespoon grated onion

⅓ teaspoon salt

Soften gelatine in 2 tablespoons cold water, then dissolve in the boiling water, but do not cook after gelatine is put in. Add all other ingredients. Mold and chill. Serve with cooked or mayonnaise salad dressing, plain or on lettuce leaves.

SALADS PROVIDE AN EASY METHOD OF USING LEFTOVERS

MIXED VEGETABLE SALAD

1 cup cooked potatoes

1 cup cooked carrots

1 cup cooked peas

1 cup cooked beets

Make a French dressing of

½ cup oil

½ teaspoon salt

2 tablespoons vinegar

⅛ teaspoon cayenne

Mix dressing thoroughly and pour over the vegetables. If vegetables are kept in different bowls instead of mixed together, the flavor of the salad is improved. Any vegetable may be used in this way. Let stand 30 minutes. When ready to serve, place each portion in a nest made of two lettuce leaves or other salad, green. If desired, cooked dressing may be mixed with the vegetable in place of French dressing, or may be served with it.

[pg 100]

EGYPTIAN SALAD

>1 cup left-over baked beans, cooked dried peas, or beans or lentils, or cooked rice, rice.
>
>1 cup chopped celery
>
>3 tablespoons chopped pepper
>
>3 tablespoons chopped pickle
>
>1 cup cooked salad dressing

Mix ingredients thoroughly and let stand 30 minutes to blend flavor thoroughly.

CABBAGE, PEANUT AND APPLE SALAD

>2 cups chopped cabbage
>
>1 cup peanuts
>
>1 cup chopped apples
>
>1 cup salad dressing

Mix ingredients and serve with French dressing. This salad looks very appetizing when served in cups made of hollowed out red apples, the pulp removed being used in the salad.

CHEESE SALAD

>1 cup American or cream cheese
>
>2 tablespoons vinegar
>
>⅓ cup oil
>
>½ teaspoon salt
>
>⅛ teaspoon cayenne

    2 tablespoons chopped olives

    3 tablespoons chopped nuts

Blend all ingredients thoroughly. Shape as desired and chill. Serve with French dressing. (If American cheese is used, grate or cut fine.)

FRUIT SALAD

Left-over small portions of fruits may be blended in almost any combination to form a salad. Plain French dressing or French dressing made with fruit juice in place of vinegar, or cooked dressing or mayonnaise may be combined with the fruit. Bananas combine well with any other fruit and, being the least expensive fruit, may be used as the basis of fruit salads.

MANDALAY SALAD

    1 cup cooked peas or carrots

    1 cup cooked cold rice

Mix with dressing made of

    ⅓ cup oil

    1 tablespoon vinegar

    ¼ teaspoon salt

    ⅛ teaspoon cayenne

    ¼ teaspoon curry powder

Mix all ingredients; serve cold, either plain, on lettuce leaves, or in nests made of cabbage or celery.

[pg 101]

POTATO SALAD

2 cups potatoes from fresh-cooked, or left-over baked, boiled or mashed potatoes.

¼ cup chopped parsley

1 teaspoon onion juice

1 cup cooked salad dressing

3 tablespoons chopped green pepper may be added if desired.

If mixed while cooked dressing is hot, then chilled, the flavor is much improved.

Left-over mashed potatoes may be combined with cooked corn and green pepper for a delicious salad.

MEAT OR FISH SALAD

1 cup left-over meat or fish

3 tablespoons chopped pickle

½ cup chopped celery

1 cup cooked salad dressing

Mix ingredients thoroughly and serve. If one-quarter cup of French dressing is mixed with meat or fish, 30 minutes before adding other ingredients, the flavor is much improved.

CAULIFLOWER SALAD

1 cup cooked cauliflower

1 cup cooked salad dressing

3 tablespoons chopped pickle

1 tablespoon chopped pimento

1 tablespoon vinegar

Blend ingredients thoroughly and serve. Cauliflower which has been creamed or scalloped may be used, if sauce is carefully rinsed from the vegetable.

CARROT SALAD

Grind raw carrot in food chopper. Make French dressing with chicken fat instead of oil. Mix ingredients and serve.

    1 cup raw carrots

    ½ cup oil (preferably oil from chicken fat)

    1 tablespoon vinegar

    ½ teaspoon salt

    1 tablespoon parsley

    ⅛ teaspoon paprika

HINDU SALAD

    2 tablespoons flour

    1 teaspoon salt

    1 egg

    ⅛ teaspoon cayenne

    2 tablespoons granulated gelatine, plus 2 tablespoons cold water

    1 teaspoon mustard

    1 teaspoon curry powder

    3 tablespoons melted fat

    1 cup milk

    ⅓ cup vinegar

    2 cups cooked rice

    2 tablespoons chopped olives

Mix dry ingredients, add egg and blend thoroughly. Add melted fat, milk and vinegar. Cook over hot water until thick as custard. Soften gelatine in cold water. Add to the hot dressing. When dissolved add rice and olives, place in mold and chill. Serve plain or with ½ cup French dressing.

[pg 102]

THE USE OF STALE BREAD, CAKE, AND LEFTOVER CEREAL

DATE CRUMB PUDDING

    1 cup dried crumbs

    1 pint hot milk

Let stand until milk is absorbed, then add

    ¼ teaspoon salt

    ½ cup molasses

    ½ teaspoon cinnamon

    1 cup dates, cut small

    1 egg

    ½ teaspoon mixed cloves, nutmeg, allspice, ginger

Mix ingredients. Bake 40 minutes in moderately hot oven. This pudding is so well flavored that it does not really require a sauce, but if one is desired the molasses sauce on page *86, or the hard or lemon sauce on page *87 will be found to suit.

FIG PUDDING

    ¼ lb suet

    ½ lb chopped figs

    1 cup sour apple (cored, pared and chopped)

1 cup milk

½ cup molasses

½ cup corn syrup

1 cup breadcrumbs

2 eggs

⅓ cup flour

Cream suet; add figs, apple and corn syrup. Pour milk over bread. Add yolks, beaten. Combine. Add flour and egg whites. Steam 4 hours.

FRUIT TAPIOCA

¼ cup pearl tapioca

⅓ cup corn syrup, or

¼ cup sugar

⅛ teaspoon salt

1 cup water

1 cup milk

1 cup fruit

Soak tapioca in the water over night. Add the other ingredients except the fruit and cook over hot water until the tapioca is clear. Add fruit and 1 teaspoon vanilla and chill.

RICE FRUIT CUSTARD

⅓ cup rice

1 cup milk

⅓ cup corn syrup

1 teaspoon vanilla

⅛ teaspoon salt

1 egg

1 cup fruit

Cook rice with milk in double boiler 30 minutes. Add other ingredients and cook 10 minutes. Chill and serve.

[pg 103]

NUT AND FRUIT PUDDING

1 cup stale breadcrumbs

2 cups scalded milk

½ cup corn syrup

½ cup chopped nuts

2 eggs

⅛ teaspoon salt

½ teaspoon vanilla

¾ cup chopped figs, dates or raisins

Pour scalded milk over breadcrumbs. Beat eggs. Add other ingredients. Bake 25 to 35 minutes in moderate oven.

CHOCOLATE BREAD PUDDING

1 cup crumbs

2 cups milk

1 oz. chocolate

⅓ cup sugar

½ cup corn syrup

2 eggs

⅛ teaspoon salt

½ teaspoon vanilla

Use whites for meringue with 2 tablespoons corn syrup.

CAKE CROQUETTES

    1 pint stale cake crumbs

    1 cup milk

Soak 1 hour; heat and add

    2 yolks of eggs

    2 teaspoons vanilla

Chill, shape, roll in eggs and crumbs and brown in frying pan. Serve with hard sauce.

CEREAL FRUIT PUDDING

    2 cups milk

    1 cup any ready-to-eat cereal

    1 egg (beaten)

    ⅓ cup molasses

    ½ teaspoon salt

    ½ teaspoon cinnamon

    1 cup raisins, dates or prunes

Mix ingredients. Bake 30 to 40 minutes in moderately hot oven.

SCALLOPED FISH

    2 cups crumbs

    2 cups fish

    2 tablespoons chopped parsley

    ¼ cup fat

¼ cup flour

⅛ teaspoon pepper

2 teaspoons onion juice

1½ cups milk

1 teaspoon salt

2 tablespoons fat

Melt fat, add dry ingredients and gradually the liquid to make a smooth sauce. Add onion juice, lemon juice, parsley and fish. Mix [pg 104] with crumbs 2 tablespoons fat. Place crumbs on top. Bake in greased pan 25 minutes.

SPANISH CASSEROLE

2 cups cooked rice

1 quart tomatoes

¼ to 1 lb. hamburg steak

⅛ teaspoon pepper

3 teaspoons salt

2 tablespoons onions, chopped

⅛ teaspoon cayenne

Add rice to tomatoes. Add seasoning and meat, browned. Bake in casserole about 2 hours.

PEANUT LOAF

3 cups stale bread crumbs

2 cups milk

2 teaspoons salt

⅛ teaspoon pepper

¼ teaspoon poultry seasoning

1 tablespoon onion juice and pulp

 2 eggs

 4 teaspoons baking powder

 1½ cups chopped peanuts

Add bread to milk; add seasoning, beaten eggs, baking powder, and peanuts. Pour into greased, lined baking tin. Bake in moderate oven 40 minutes.

CHEESE ENTREE

 1 cup cooked farina or rice

 1 cup cheese

 1 cup nuts

 1 cup milk

 ⅛ teaspoon cayenne

 1 egg

 1 teaspoon salt

Mix ingredients thoroughly. Bake in greased dish 30 minutes.

BEAN LOAF

 2 cups cold cooked beans

 1 egg beaten

 1 cup breadcrumbs

 ⅛ teaspoon pepper

 1 tablespoon minced onion

 2 tablespoons catsup

 ¼ teaspoon salt

Shape into loaf. Bake 25 minutes. Serve with tomato sauce.

ROYAL FRENCH TOAST

Use leftover bread as French toast by dipping in mixture of

1 cup milk

1 tablespoon corn syrup

1 egg beaten

Then brown in frying pan in small amount of fat. Spread with marmalade, jelly, cocoanut, or preserves and serve as dessert.

DRIED FRUIT PUDDING

One cup dried apricots, peaches or prunes soaked two hours in two cups of water.

1 cup bread crumbs

⅔ cup corn syrup

1 teaspoon orange or lemon rind

2 eggs

⅛ teaspoon salt

1 teaspoon lemon juice

½ cup chopped nuts

Mix ingredients. Place in greased baking dish and bake 30 minutes in moderately hot oven.

CHEESE SAUCE ON BREAD

¼ cup fat

1 pint milk

2 qts. milk

¼ cup flour

¼ teaspoon cayenne

1 cup cheese

Make as white sauce and add cheese. Pour over bread, sliced and toasted. Bake in moderate oven.

SURPRISE CEREAL

3 cups dried breadcrumbs

3 tablespoons maple syrup

½ teaspoon salt

Mix thoroughly and place in moderately hot oven for 20 minutes, stirring frequently. Remove and serve as breakfast food. Very inexpensive and delicious. Graham, corn or oatmeal bread is best for this purpose, but any bread may be used.

SURPRISE CROQUETTES

1 cup leftover cereal

1 cup chopped peanuts

½ cup dried breadcrumbs

1 beaten egg

Shape as croquettes and bake in oven or pan-broil. Serve with tart jelly.

CHEESE STRAWS

1 cup stale bread

⅛ teaspoon cayenne

½ cup grated cheese

¼ cup milk

⅔ cup flour

¼ teaspoon salt

Make into dough; roll ¼ inch thick. Cut into strips 6 inches long and ½ inch wide. Place on baking sheet. Bake 20 minutes in moderate oven. Serve with soup, salad, or pastry.

[pg 106]

SOUPS UTILIZE LEFTOVERS

In nearly every case when meat is purchased, some bone is paid for. Too frequently this is either left at the market or thrown away in the home. Bones, gristle, tough ends, head and feet of chickens, head, fins and bones of fish, etc., should be utilized for making soup.

If a meat or fish chowder with plenty of vegetable accompaniment is served, no other meat is required for the usual home meal.

If a cream of dried or fresh vegetables, or a meat stock soup with plenty of vegetables or cereal content, is served, the amount of meat eaten with the main course of the meal will be materially lessened.

Soups may be a most economical method of using water in which meat, fish or vegetables have been cooked; also of utilizing small portions of leftover meats, fish, vegetables or cereal.

Cream soups are made by cooking vegetables or cereal, then utilizing the water in which they are cooked as part of the liquid for the soup. Outer parts or wilted parts of vegetables may be utilized for soups instead of being discarded. Water in which ham or mutton has been boiled makes an excellent basis for dried or fresh vegetable soups. In fact, soup can be made from all kinds of leftovers—the variety and kind make little difference so long as the mixture is allowed to simmer for several hours and is properly seasoned.

CREAM SOUP

⅓ cup fat

⅓ cup flour

1 teaspoon salt

1 cup cereal or vegetable

¼ teaspoon cayenne

1 pt. milk

1 pt. water, in which vegetable or cereal was cooked, or leftover water in which meat was cooked.

Melt fat, add dry ingredients and, gradually, liquid. When at boiling point, add vegetables or cereal and serve.

MEAT STOCK

Leftover bits of meat, bone, or gristle may be used alone or with some fresh meat and bone from shin or neck.

To each 1 lb. of meat and bone, add 1 qt. cold water. Let stand 1 hour. Cover and bring slowly to boiling point and simmer 2 to 3 hours. Remove bones and meat. Let stand until cold. Skim off fat. Add vegetables cut in small pieces, season as desired and cook until vegetables are tender. Leftover cereals, as barley, oatmeal, etc., vegetables, macaroni, tapioca, sago, etc., etc., may be added for increased food value.

[pg 107]

TOMATO GUMBO SOUP

Bones and gristle from chicken or turkey

2 qts. cold water

1 cup okra

1 tablespoon chopped pimento

1½ teaspoons salt

½ cup rice

2 tablespoons fat

1½ cups tomatoes

¼ cup chopped parsley

Soak bones and gristle in the cold water 1 hour. Then boil slowly 1 hour, in same water. Strain out the bones and gristle and add other ingredients to the liquor. Boil this mixture slowly ¾ hour and serve.

LEGUME SOUP

1 cup dried peas, beans or lentils

3 qts. cold water

1 tablespoon onion pulp

1 ham bone or ½ pound smoked sausage

1 teaspoon celery salt

2 teaspoons salt

2 tablespoons flour, plus

2 tablespoons cold water

¼ teaspoon pepper

1 cup tomato

Wash and soak dried legume over night. In morning drain, add water, ham bone or sausage and cook very slowly until tender. Add other ingredients, cook ½ hour and serve.

VEGETABLE SOUP

1 qt. boiling water

½ cup carrots

½ cup cabbage

1 cup potatoes

1 cup tomato juice and pulp

1 tablespoon minced onion

¼ teaspoon pepper

4 tablespoons fat

4 cloves

1 bayleaf

2 teaspoons salt

4 peppercorns

2 tablespoons chopped parsley

Heat onion, pepper, salt, bayleaf and peppercorns with tomatoes for 20 minutes. Strain. To juice and pulp add other ingredients and cook slowly 1 hour. Add parsley just before serving.

CREAM OF CARROT SOUP

2 cups diced carrots

2 cups water

1 cup milk

⅛ teaspoon pepper

2 tablespoons fat

2 tablespoons flour

1 teaspoon salt

Cook the carrots in the water until tender. Melt the fat, add dry ingredients, add gradually the 1 cup water in which the carrots were cooked and the milk. When at boiling point, serve with a little grated [pg 108] raw carrot sprinkled over top of soup. Any vegetable, raw or cooked, may be used in the same way, as cauliflower, cabbage, peas, turnips, etc.

SALMON CHOWDER

1 cup cooked or canned fish

1 cup cooked potato, diced

1 cup peas

2 tablespoons fat

2 tablespoons flour

1½ teaspoons salt

¼ teaspoon paprika

2 cups milk

1 cup water from boiled potatoes

2 tablespoons chopped parsley

1 teaspoon onion juice

Melt fat, add dry ingredients and gradually the liquid. When at boiling point, add parsley and serve.

CHEESE CREAM SOUP

1 cup cheese

2 cups milk

2 tablespoons fat

1¼ teaspoons salt

¼ teaspoon cayenne

½ teaspoon celery salt

3 tablespoons flour

Melt fat, add dry ingredients and gradually the liquid. When at boiling point and just ready to serve add cheese. Any kind of cheese may be used for this purpose.

BEAN SOUP

1 cup beans

1 quart water

1 tablespoon onion juice

¼ teaspoon Worcestershire sauce

1 cup brown stock

¼ teaspoon celery salt

2 teaspoons salt

¼ teaspoon cayenne

1 hard cooked egg

1 lemon, sliced

¼ teaspoon mustard

2 tablespoons flour, plus 2 tablespoons cold water

Soak beans over night, drain. Place in 1 quart of fresh cold water and cook until very tender. Add other ingredients and bring to boiling point. Slice thin, hard cooked egg and lemon from which seeds have been removed and serve with each portion. Do not remove lemon rind as this gives a piquant flavor.

POTATO AND CHEESE SOUP

2 cups cooked diced potatoes

2 cups water in which potatoes were cooked

1 cup milk

2 teaspoons onion juice

2 tablespoons fat

3 tablespoons flour

1½ teaspoons salt

⅛ teaspoon cayenne

2 tablespoons of finely chopped parsley

¼ cup grated cheese

Dice potatoes and cook slowly until very tender. Rub through strainer, using potato and 2 cups of the water. Melt fat, add dry ingredients and gradually the liquids and onion juice. When ready to serve, sprinkle parsley and cheese over top.

[pg 109]

ALL-IN-ONE-DISH MEALS

NEED ONLY FRUIT OR SIMPLE DESSERT, AND BREAD AND BUTTER TO COMPLETE A WELL-BALANCED MENU

LENTILS WITH RICE AND TOMATOES

¾ cup lentils

1 cup rice

1 quart tomatoes

1 teaspoon Worcestershire

2 teaspoons salt

¼ teaspoon cayenne

¼ teaspoon bay leaf

¼ teaspoon sage

Soak lentils over night; drain; add one quart fresh water and one teaspoon of salt. Cook slowly until tender. Add other ingredients. Steam or bake for 45 minutes.

RICE, TOMATOES, GREEN PEPPER AND BEEF

½ cup cooked rice

1 pint tomatoes

⅓ cup green pepper chopped

2 cups fresh or left-over cooked meat

2 teaspoons salt

¼ teaspoon cayenne

Mix all ingredients. Bake in greased dish slowly for one hour.

HOMINY AND CURRIED MUTTON WITH BEETS

1 cup hominy which has been soaked over night, drained

1 quart fresh water and 1 teaspoon of salt added; cook until tender

2 cups mutton from shoulder

1 teaspoon kitchen bouquet

1 teaspoon curry

2 cups water

1 teaspoon Worcestershire sauce

1 tablespoon cornstarch

1 cup diced beets

1 teaspoon salt

⅛ teaspoon cayenne

Mix all ingredients thoroughly. Bake in covered casserole slowly for one hour. Mutton should be cut in about one-inch pieces.

TAMALE PIE MADE WITH CORNMEAL MUSH, MEAT AND CHOPPED PEPPERS

4 cups water

1 cup cornmeal

2 teaspoons salt

⅓ cup chopped peppers

2 cups cooked meat

⅛ teaspoon cayenne

To cornmeal add one-half cup of cold water. Boil three cups of water and add cornmeal. Boil five minutes. Add other ingredients. Cook in greased baking dish for one hour.

[pg 110]

BAKED SOY BEANS WITH GREENS AND TOMATO

    1 pint soy beans

    ¼ lb. salt pork

    ½ teaspoon soda

    ⅛ teaspoon cayenne

    1 onion

    1½ tablespoons salt

    ¾ cup molasses

    ¾ tablespoon mustard

    Boiling water (about one quart)

    1 pint tomatoes

    2 cups cooked spinach

Soak beans over night; drain. Cover with fresh water and the soda and boil, until skins break, but do not let beans become broken. Cut rind from salt pork and cut into six or eight pieces. To 1 cup of boiling water add the cayenne, salt, molasses, mustard and tomatoes. In bottom of bean pot place the onion and a piece of salt pork. Add beans. Pour over this the seasonings. Cover the beans with boiling water. Bake three hours covered. Uncover, put spinach to which has been added 1 teaspoon of salt, 1 tablespoon of vinegar, one-eighth teaspoon of pepper, on top. Bake 30 minutes and serve.

CASSEROLE OF KIDNEY BEANS, SALT PORK AND SPINACH

One cup of kidney beans, soak over night; drain. Cover with fresh water. Add 2 teaspoons of salt, cook in small amount of water until tender. Force through colander. Measure 1½ cups and add one-quarter pound salt pork chopped fine, 1 teaspoon Worcestershire sauce, 1 cup of water or meat stock or gravy.

Place half of mixture in greased baking dish. Cover with two cups of spinach, to which has been added one-quarter cup of vinegar, 2 tablespoons of fat and one-half teaspoon of salt. Cover with other half of bean mixture. Bake 20 minutes.

SCALLOPED MACARONI WITH PEAS IN TOMATO AND CHEESE SAUCE

    1 cup macaroni

    1 cup peas

    1 pint tomatoes, juice and pulp

    1 cup grated cheese

    ¼ cup fat

    ¼ cup flour

    1 teaspoon salt

    ⅛ teaspoon cayenne

Cook macaroni until tender in one quart of boiling water and one teaspoon of salt; drain. Melt fat, add flour, salt and cayenne. Gradually add tomatoes and when at boiling point remove from fire, add cheese and peas. Place macaroni in greased baking dish, pour sauce over it and bake 30 minutes.

[pg 111]

CURRIED RICE WITH CORN AND CHEESE IN BROWN SAUCE

    ½ cup rice

    1 cup cheese

    1 cup corn

    1½ cup milk

    ¼ cup fat

¼ cup flour

1 teaspoon salt

⅛ teaspoon cayenne

Melt fat until brown. Add flour and seasonings. Heat until brown. Add milk gradually. When at boiling point add other ingredients. Place in baking dish and bake 45 minutes.

FISH AND VEGETABLE CHOWDER

3 lbs. fish

2 cups diced potatoes

⅓ cup chopped onion

½ cup chopped salt pork

1 teaspoon salt

⅛ teaspoon cayenne

1 cup peas

2 cups cold water

2 tablespoons fat

2 tablespoons flour

1 cup diced carrots

1 pint scalded milk

Cut fish into small pieces. Cover bones, fins and head with cold water. Simmer 15 minutes; strain. Cook onion and salt pork until brown. In kettle place layers of fish and mixed vegetables. To water in which bones, etc., have been cooked, add the seasonings. Mix all ingredients. Cook forty minutes, slowly, covered.

SAMP, FINAN HADDIE WITH HORSERADISH AND TOMATOES

1 smoked haddock

1 cup samp, which has been soaked over night and cooked until tender

1 quart water and 1 teaspoon of salt

2 teaspoons horseradish (grated)

1 pint tomatoes

1 teaspoon salt

¼ teaspoon cayenne

2 tablespoons cornstarch

Pour 1 cup of boiling water and one-half cup of boiling milk over fish. Let stand one-half hour, pour off liquid. Place fish in baking dish. Place samp on fish. Mix other ingredients and pour on top. Cover and bake three-quarters of an hour.

CASSEROLE OF SPAGHETTI AND CARROTS WITH PEANUTS, IN BROWN SAUCE

1 cup cooked spaghetti

2 cups brown stock

2 cups water, or

2 bouillon cubes

2 tablespoons flour

2 teaspoons salt

½ cup chopped peanuts

1 cup diced carrots

3 tablespoons chopped olives

Blend flour with 2 tablespoons cold water. Dissolve bouillon cubes in the boiling water. Mix all ingredients. Place in casserole and bake 45 minutes or until spaghetti is tender.

[pg 112]

LENTIL, PEANUT AND CHEESE ROAST WITH WHITE SAUCE AND OLIVES

    1 cup cooked lentils

    1 cup chopped peanuts

    1 cup grated cheese

    1 cup bread crumbs

    1 tablespoon fat

    2 tablespoons lemon juice

    ½ teaspoon salt

    ⅛ teaspoon cayenne

    1 teaspoon onion juice

Mix all. Place in a greased dish. Bake 30 minutes. Then pour over top a sauce made by melting 2 tablespoons of fat, adding 2 tablespoons flour, one-half teaspoon of salt and one-eighth teaspoon cayenne. Then add 1 cup of milk gradually. When at boiling point add 3 tablespoons of chopped olives. Pour this sauce over the roast and bake 20 minutes. Serve at once.

CASSEROLE OF CODFISH, PIMENTO AND CORNMEAL MUSH

    1 lb. codfish

    ⅓ cup pimento

    1 cup cornmeal

    2 cups tomatoes, juice and pulp

    2 teaspoons salt

    ⅛ teaspoon cayenne

    3 cups boiling water

Mix cornmeal with one-half cup of cold water. Add to the boiling water. Boil five minutes. In greased baking dish place fish which has been soaked over night. Place pimento on fish. Place cornmeal on pimento. To tomatoes add seasonings and pour over all. Bake slowly 45 minutes.

CURRIED VEGETABLES

One-half cup dried peas, beans or lentils, soaked over night and cooked until tender.

> ½ cup turnips
>
> ½ cup of carrots
>
> 1 cup outer parts of celery
>
> ½ cup of peas
>
> ½ teaspoon celery salt
>
> ⅛ teaspoon pepper
>
> 3 tablespoons drippings
>
> 3 tablespoons whole wheat flour
>
> 1 teaspoon curry powder
>
> 1 teaspoon salt
>
> ½ cup meat stock or water
>
> 1 cup tomato juice and pulp
>
> 1 teaspoon onion juice

Melt the fat. Add the seasoning; gradually the liquid. Add the vegetables. Cook 20 minutes. Serve very hot. This is an especially good way of adding the necessary flavor to lentils.

[pg 113]

WHEATLESS DAY MENUS

1

BREAKFAST

    Stewed Prunes

    Oatmeal

    Corn Muffins

    Top Milk

    Coffee

LUNCHEON OR SUPPER

    Cream of Spinach Soup

    All Rye Rolls

    Scalloped Potatoes

    Marmalade

DINNER

    Pot Roast

    Buttered Beets

    Fried Egg Plant

    Southern Spoon Bread

    Maple Cornstarch Pudding

2

BREAKFAST

    Dried Apricots

    Cornflakes

    Rye and Peanut Muffins

    Top Milk

    Coffee

LUNCHEON OR SUPPER
>Nut and Bean Loaf with White Sauce
>
>Corn Pone
>
>Oatmeal Cookies
>
>Currant or Plum Jelly
>
>Tea

DINNER
>Beef Casserole
>
>Baked Potatoes
>
>Green Beans
>
>Barley Biscuits
>
>Cranberry Tapioca Pudding

3

BREAKFAST
>Baked Apple Stuffed with Nuts
>
>Fried Cornmeal Mush
>
>Maple Syrup
>
>Coffee

[pg 114]

LUNCHEON OR SUPPER
>Split Pea Soup
>
>Rye Muffins
>
>Corn Oysters

Cranberry Jelly

DINNER

    Mutton Pie

    Glazed Sweet Potatoes

    Pickled Beets

    Oatmeal Bread

    Scalloped Tomatoes

    Brown Betty

4

BREAKFAST

    Dried Peaches with Jelly Garnish

    Corn Puffs and Dates

    Top of Milk

    Rye Muffins

    Coffee

LUNCHEON OR SUPPER

    Macaroni and cheese

    Corn and Rice Muffins

    Canned Fruit

    Cocoa

DINNER

    Cream of Carrot Soup

    Swiss Steak

Stewed Tomatoes
Natural Rice
Cole Slaw
Oatmeal Rolls
Brown Betty

5

BREAKFAST

Baked Apples with Marmalade Center
Cream of Grits Cereal
Top of Milk
Rye Finger Rolls
Coffee

LUNCHEON OR SUPPER

Cream of Lentil Soup
Corn Muffins
Prunes
Hot Tea

[pg 115]

DINNER

Casserole of Beef and Rice
Baked Potatoes
Stewed Corn
Cabbage Salad

Chocolate Cornstarch Pudding

MEATLESS DAY MENUS

1

BREAKFAST
    Baked Pears with Cloves and Ginger
    Cornmeal and Farina Cereal
    Coffee
    Toast

LUNCHEON OR SUPPER
    Welsh Rarebit
    Hot Tea
    Fruit Muffins
    Lettuce Salad

DINNER
    Cream of Corn Soup
    Baked Fish
    Macaroni with Tomato Sauce
    Whole Wheat Bread
    Lyonnaise Potatoes
    Orange Sago Custard

2

BREAKFAST

    Dried Peaches

    Fried Hominy

    Marmalade

    Coffee

    Popovers

LUNCHEON OR SUPPER

    Bean Soup

    Lettuce Salad

    Cheese Straws

    Olives

[pg 116]

DINNER

    Chicken Fricassee

    Dumplings

    Baked Squash

    Peas

    Cranberry Jelly

    Barley Muffins

    Mock Mince Pie

3

BREAKFAST

Oranges
Pearled Barley
Top Milk
Currant Jelly
Rye Bread Toasted
Coffee

LUNCHEON OR SUPPER

Mixed Vegetable Salad
Boston Brown Bread
Hot Tea

DINNER

Clam Chowder
Spinach and Cheese Loaf
Carrots
Creamed Cauliflower
Oatmeal Nut Bread
Spice Pudding
Hard Sauce

MEAT SUBSTITUTE DINNERS

Consommé with Spaghetti
Cornmeal Muffins
Cabbage and Cheese
Julienne Potatoes
Carrots

Dressed Lettuce

Jellied Prunes with Nuts

Thin Bean Soup

Rye Rolls

Corn and Oyster Fritters

Baked Potato

Scalloped Tomato

Apple and Celery Salad

Graham Pudding with Hard Sauce

[pg 117]

Consommé with Tapioca

Brown Bread

Salmon Loaf or Escalloped Salmon

Creamed Potatoes

Peas

Lettuce Salad

Gelatine Dessert

Thin Cream of Celery Soup

Rye Bread

Nut Loaf

Brown Sauce

Scalloped Potatoes

Spinach

Lettuce Salad with Tomato Jelly

Sago Pudding

Scalloped Hominy and Cheese

Swiss Chard or Spinach

Whole Wheat Bread

Stuffed Baked Potato

Baked Pears

Molasses Cookies

Escalloped Codfish

Baked Onions

Corn Bread

Apple Salad

Fig and Date Pudding with Tart Jelly

Cream of Barley Soup

Turkish Pilaf

War Muffins

Apple and Cabbage Salad

Chocolate Bread Pudding

Cream of Rice Soup

Rye Meal Rolls

Kidney Bean Croquette

Greens

Dried Apricot Butter

Oranges, Bananas and Dates

Ginger Cookies

Bean Soup

Welsh Rarebit or a Cheese Dish

Natural Rice

Tomato Sauce

Corn Meal Parker House Rolls

Dried Peach Pudding

[pg 118]

VEGETABLE DINNERS

- Corn Soup
- Oatmeal Bread
- Nut Loaf
- Tomato Sauce
- Green Beans
- Potatoes au Gratin
- Jellied Prunes
- Boston Roast
- Tart Jelly
- Whole Wheat Bread
- Creamed Cauliflower
- Squash
- Cranberry Slump
- Kidney Beans with Rice
- Fried Apples with Raisins
- Celery in Brown Sauce
- Cornmeal Baking Powder Biscuits
- Tapioca Cream
- Baked Beans
- Boston Brown Bread
- Spinach
- Apple and Pimento Salad
- Gelatine Dessert
- Cream of Vegetable Soup
- Lima Bean Croquets

Creamed Potatoes

Carrots

Pickled Beets

Cornmeal and Rye Muffins

Cottage Pudding

Cream of Celery Soup

Rye Bread

Spinach Loaf

Cabbage and Pepper Relish

Brown Rice

Marmalade Pudding

Cream of Tomato Soup

Corn Sticks

Baked Macaroni and Cheese

Baked Sweet Potatoes

Eggplant

Beet and Cabbage Relish

Whole Wheat Bread

Apricot Shortcake

Hard Sauce

[pg 119]
　Of our men we ask their lives; Of ourselves, a little less food.

SAVE AND SERVE

TO SAVE BREAD. Serve bread or rolls made from corn, rye or from coarse flours. Use breakfast foods and hot cakes, composed of corn, oatmeal, buckwheat, rice or hominy. Serve no toast as garniture or under meat. Serve war breads. Use every part of the bread, either fresh or stale, for puddings and toast; or dried and sifted for baked croquettes; or use to extend flour in the making of muffins and drop cakes.

TO SAVE MEAT. Use more chicken, hare, rabbits, duck, goose, lobster, oysters, clams and egg and cheese dishes of all kinds. Use less beef, mutton, and pork and serve smaller portions at table of these meats. Have fewer of these items on the menu. Provide more entrees and made-over dishes in which a smaller quantity of meat is extended by the use of potatoes, rice, hominy, etc. Use beans, as they contain nearly the same nutritive value as meat. Serve bacon only as a dish and not as a garniture, and this way not more than once a week. Use cheese, dried vegetables and nuts. Use fish and meat chowders. Use meat extension dishes. Serve vegetable dinners.

TO SAVE SUGAR. Use less candy and sweet drinks. Use honey, maple sugar, corn syrup, molasses and dark syrups with hot cakes and waffles and in all cooking, in order to save butter and sugar. Use all classes of fruit preserves, jam, marmalades and jellies. Do not frost or ice cakes. Serve dried fruits with cereals, and no sugar is needed.

[pg 121]

TO SAVE FATS. Serve as few fried dishes as possible, so as to save both butter and lard, and in any event use vegetable oils for frying—that is, olive oil, corn oil, cottonseed oil, vegetable oil compounds, etc. Trim all coarse fats from meats before cooking and use the waste fats for shortening and for soap. We are short of soap fats as our supplies of tropical oils used for soap-making are reduced. Do not waste soap. Save fat from soup stock and from boiled meats. Use butter substitutes where possible.

TO SAVE MILK. Use it all. Buy whole milk and let cream rise. Use this cream, and you secure your milk without cost. Economize on milk and cream except for children. Serve buttermilk. Serve cottage cheese regularly in varying forms. It is especially nutritious. Use skimmed milk in cooking. A great quantity of it goes to waste in this country. Use cheese generally. The children must have milk whole, therefore reduce the use of cream.

<u>USE VEGETABLES.</u> Use more vegetables and potatoes. Make fruits and vegetables into salads and attractive dishes. Feature vegetable dinners and salads of all kinds. Encourage the use of cheese with salads. Make all types of salads from vegetables. We have a great surplus of vegetables, and they can be used by substituting them for staples so that the staples most needed will be saved.

Make all kinds of vegetable soups, especially the cream soups, in which the waste from staple vegetables, such as outer leaves and wilted parts, can be utilized. These are wholesome and nutritious and save meat.

[pg 122]

www.ingramcontent.com/pod-product-compliance
Lightning Source LLC
Chambersburg PA
CBHW031624210526
45464CB00004B/1731